Water, Place, and Equity

American and Comparative Environmental Policy

Sheldon Kamieniecki and Michael E. Kraft, series editors

Water, Place, and Equity

edited by John M. Whiteley, Helen Ingram,
and Richard Perry

The MIT Press
Cambridge, Massachusetts
London, England

For information about special quantity discounts, please e-mail special_sales@ mitpress.mit.edu

This book was set in Sabon by Graphic Composition, Inc., Bogart, Georgia. Printed on recycled paper and bound in the United States of America.

Library of Congress Cataloging-in-Publication Data

Water, place, and equity / edited by John M. Whiteley, Helen Ingram, and Richard Warren Perry.
    p.   cm. — (American and comparative environmental policy)
Includes bibliographical references and index.
ISBN 978-0-262-23271-5 (hardcover : alk. paper) — ISBN 978-0-262-73191-1 (pbk.)
1. Water resources development. 2. Equity. I. Whiteley, John M., 1940–
II. Ingram, Helen M., 1937– III. Perry, Richard Warren.
HD1691.W325 2008
333.91—dc22
2008017001

10 9 8 7 6 5 4 3 2 1

Dedication

This book is dedicated to the late Chris Nunn Garcia. The mother of three children, she graduated with distinction with a Bachelor of Arts degree from the University of New Mexico and earned a Ph.D. in natural resources economics and law from the University of Wisconsin, Madison.

Chris worked in academics, public policy research, and community service, pursuing each with her characteristic charm and single-minded devotion. Her involvement in one project prompted a colleague to remark, "Chris was recorder, institutional memory, and conscience." She was an award-winning professor and highly regarded multidisciplinary

academic. Chris's many community projects include the founding of the New Mexico Regional Water Planning Dialogue. Also, she headed several major projects for the University of New Mexico Law School's Utton Center for Transboundary Resources.

Chris was inquisitive and threw herself into everything she touched. This was true in her professional work as well as her friendships. She appreciated the value in everyone's unique position. Her genuine interest in the perspectives of others is why so many people felt that Chris understood them and why her work was often much larger than herself.

The commitment of the contributors of this book to interdisciplinarity inquiry and more equitable public policy is a commitment reflected in Chris's life's work.

Contents

Series Foreword

There is little question that the availability and distribution of global water resources rank among the most profound environmental policy challenges of the twenty-first century. Water is a scarce resource today in many parts of the United States as well as worldwide, and it is likely to become even scarcer in the years ahead. Among the major reasons are substantial growth in water-intensive economic activities, especially in arid or semi-arid areas; the effects that climate change is likely to have on precipitation rates and evaporation, which will compound the problem of development pressures; and a continuing increase in the world's population that will add perhaps another three billion people to Earth's total by 2050.

Highly uneven and often inequitable access to vital water resources is striking today. This may well lead to serious and potentially more disruptive social, economic, and political conflicts in the future. So it is no exaggeration to say, as the authors of this volume argue, that water is likely to dominate natural resource politics around the world in the twenty-first century as much as oil influenced global politics in the late twentieth century. Whether the issue is one of potential conflict between water-rich areas of the United States and the arid Southwest and West, how to ensure the sustainability of groundwater in the Middle East, or how to improve the efficiency of agricultural water use worldwide, one conclusion is certain. The coming decades will require fresh perspectives on water resources that have long been taken for granted and reconsideration of the most appropriate principles of water resource governance and public policies.

In this innovative volume, the editors and contributors make a strong case for the primacy of equity considerations in the distribution of water resources. This perspective challenges the conventional view that norms of efficiency should govern water use, and that markets can be relied upon to distribute water as long as the proper pricing mechanisms are in place. As many of the cases discussed in the book make clear, human uses of water may conflict with ecological or cultural needs in ways that are not easily amenable to resolution. Nor can water supplies be increased simply by improving infrastructure. Sooner or later, new strategies of public involvement and stakeholder collaboration, among others, will seemingly become essential. At that point the global quest for sustainable water resource governance may depend upon how well equity concerns can be integrated with the long-standing centrality of utilitarian management of this precious resource. In this regard, this book makes an original and important contribution to the literature on water policy.

By focusing on water use in specific places and both older and newer perspectives on equity, the contributors present an array of intriguing normative case studies. These cases help to build understanding of how equity concerns arise and the effects they can have on decisions over water allocation. The authors' analyses of the cases also suggest promising paths for future research, and thus help to ensure that others can build on the distinctive scholarly and practical policy contributions in this collection.

This book illustrates well our purpose in the MIT Press book series American and Comparative Environmental Policy. We encourage work that examines a broad range of environmental policy issues. We are particularly interested in volumes that incorporate interdisciplinary research and focus on the linkages between public policy and environmental problems and issues both within the United States and in cross-national settings. We welcome contributions that analyze the policy dimensions of relationships between humans and the environment from either a theoretical or empirical perspective. At a time when environmental policies are increasingly seen as controversial and new approaches are being implemented widely, we especially encourage studies that assess policy successes and failures, evaluate new institutional arrangements and policy tools, and clarify new directions for environmental politics and policy. The books in this series are written for a wide audience that includes academics, policymakers, environmental scientists and professionals, business and labor leaders, environmental activists, and students concerned with environmental issues. We hope they contribute to public understanding of environmental problems, issues, and policies of concern today and also suggest promising actions for the future.

Sheldon Kamieniecki, *University of California, Santa Cruz*
Michael Kraft, *University of Wisconsin-Green Bay*
American and Comparative Environmental Policy Series Editors

Acknowledgments

This book grew out of a conference on "Challenges of a Transboundary World" held at the Beckman Center of the National Academies of Sciences and Engineering in 2004 with broad sponsorship within the University of California. Financial support is gratefully acknowledged from the following organizations and their leaders: the University of California, Institute on Global Conflict and Cooperation, and its International Environmental Policy Research Program; the Focused Research Group in International Environmental Cooperation and its director, Professor Joseph F. DiMento; the Program in Industrial Ecology and its director, Professor Oladele Ogunseitan; the Center for Global Peace and Conflict Studies and its director, Professor Wayne Sandholtz; the Center for the Study of Democracy and its director, Professor Russell Dalton; the School of Social Sciences and its dean, Professor Barbara Dosher; and the School of Social Ecology and its dean, Professor C. Ronald Huff. The conference itself honored Professor Helen Ingram for the significance of her personal, positive impact on the lives of emerging scholars, and her academic mentorship.

Three individuals deserve special recognition for their dedication to the strengthening of this manuscript: Kay Collins, a specialist in government information in the UCI Libraries, once said that new electronic digital sources allowed her to contribute to scholarship by helping find information she did not know existed in places she would not have thought to look. Dianna Sahhar, a specialist in document delivery services from the UCI Libraries, is very resourceful in finding new ways to acquire essential documents, especially those that are hard to locate because they predate the digital era. She has a demanding set of responsibilities but always scheduled time to help. Norma Yokota, an analyst on complex projects within the School of Social Ecology, is technical editor for this volume and a most resourceful colleague in getting all of the contributors onto "the same page." This heartfelt appreciation extends to her efforts in the organization and administration of the conference in 2004, which inspired this volume.

John M. Whiteley
Irvine, California

Water, Place, and Equity

1

The Importance of Equity and the Limits of Efficiency in Water Resources

Helen Ingram, John M. Whiteley, and Richard Perry

Importance of Water

Water will dominate world natural resource politics by the end of the twenty-first century much as oil dominated the late twentieth century. Even more profoundly than oil before it, water permeates the larger part of political, economic, social, and even religious conflicts. Water is also integral to a sense of place, and fairness in its allocation will be a fundamental cornerstone to a more equitable future for humankind.

No longer can water be found in the quantity and the quality, and at a price, that can accommodate the ever-accelerating demand for it around the world. Though it may well be the case that the total volume of water on the planet is sufficient for societies' needs, the largest portion of this water is located in the wrong places, and it is misallocated, wasted, or degraded by pollution. The poorest of the poor people, perhaps more than 800 million, live in arid areas and depend on water for life and livelihood (UN 1999, 2004). The general anxiety about global warming—accompanied by specific panics about flooded cities like New Orleans, or about advancing deserts in the Sahel region of Africa, or about the melting ice sheets at both of the earth's poles—has simply heightened the stakes of a problem that has been long in the making.

A growing recognition of the importance of water is marked by the fact that access to water is widely regarded as a basic human right. The United Nations articulated in 1992 the right of all human beings to have access to clean water and sanitation at an affordable price (UN 1992). The South African Constitution in its Bill of Rights has gone so far as to guarantee to every citizen access to sufficient water for life, and to make the provision of life-sustaining water the positive obligation of government (Stein 2000). The understanding of water as a human right derives from the nature of our water-dependent and water- suffused bodies and from the nature of our waterborne civilizations. Yet, all signs are that the water crisis is getting worse and will continue to do so unless corrective action is taken (UN/WWAP 2003).

Insufficient and degraded waters are tracers pointing directly to problems of global poverty and inequality. Water and sanitation have a prominent place in the Millennium Development Goals of the United Nations that set targets to halve the 1.1 billion people currently without access to drinking-water services by 2015 and the 2.2 billion without proper sanitation by 2020 (Schouten and Schwartz 2006). So far, the levels of investment are falling woefully short, and even if greater funding were provided, there remain serious issues of water governance. The 2003 United Nations Report *Water for People, Water for Life* concludes, "This crisis is one of water governance, essentially caused by the ways we mismanage water" (UN/WWAP 2003, 4).

The movement for water governance reform began nearly two decades ago. The tenets for improved governance are endorsed by nearly all water academics, professionals, and institutions including the World Bank. These tenets for improved governance include full-cost recovery water pricing, the reduction of subsidies, water transfers through markets, and decentralization of control and citizen participation in watershed management and in the allocation and distribution of public supply. Yet, water continues to be an enormously contentious issue, and to date the fruits of governance reform fall far short of the promise of benefit to humankind, especially to the poorest of the poor.

One of the major impediments to attaining higher levels of agreement on water management is the failure to recognize value pluralism in relation to water. In an earlier volume in this series—in which the editors and several authors of the present volume joined—we argued that an adequate understanding of transboundary water conflicts required a recognition of the multiple and incommensurable meanings of water in all of its specific geographical and historical sites of encounter. In that volume we argued that "water is good to think with" (Blatter and Ingram 2001, 297). In this volume we and our collaborators continue to use water to think with. More specifically, we shall use water to consider the broader notions of fairness, equity, and justice that must temper the dominant sociopolitical notions of efficiency and markets that distinguish the early twenty-first century.

This chapter will examine the historical roots and contemporary emergence of the commitment to equity and justice in water management. In the first of two sections of the book, we provide more substance to our ongoing discussion of the efficiency turn in contemporary water discourse among water scholars and practitioners. Our argument is that such reforms are needed urgently, but unless they are embedded in considerations of equity and justice, they cannot succeed. The experience of the twentieth century demonstrates unequivocally that water management based on only one or a narrow range of principles leads to a kind of "tyranny of the water commons," and its governance, or, to use a phrase from Thompson (2000), "tragically difficult."

Next, this chapter examines the case for equity in water. It recognizes equity as a protean, paradoxical concept. Competing equities previously have tended to be ignored in the literature cojoining environmental justice and water. Any articulation of the role of equity must recognize the complexity of the concept, and also recognize that if equity is to emerge, it must do so in specific places under particular circumstances—there is no "one size fits all" conception of equity that is workable. The interrelationships of arguments made in this chapter draw directly on other chapters in this book and are cited throughout. The last section of the chapter reflects on the importance of equity to the future of water governance and principles of water allocation for the remaining years of the twenty-first century. It provides a broad overview of both parts of the book.

The Efficiency Framework in Contemporary Water Politics

The conceptual frame for resolving water disputes on which much of contemporary academic and political analysis has settled is a focus on increased efficiency. It is now widely agreed that simply continuing to construct new water-supply infrastructure is problematic for a host of reasons, most notably because of financial cost and adverse environmental consequences. Instead, there is a turn toward the direction of more efficient use of available water supplies, accompanied by economic incentives and market transaction mechanisms to encourage the movement of water from lower- to higher-valued uses, such the movement of water from agricultural to urban and industrial uses.

Economic efficiency in the allocation of water resources is allocation in such a manner that no further reallocation is possible that would provide gains in production or consumer satisfaction to some water users without simultaneously imposing losses on others. The efficiency framework, like other frameworks, is based on values. Individual preferences count. Societal welfare is based on the aggregation of individual choices in pursuit of individual interests. A change that makes everybody better off without making anyone worse off is a positive change in total societal welfare. Markets are the preferred institutions for pursuing economic efficiency, because there would not be a willing seller and a willing buyer unless both sides believed they were going to be better off (Young 2005). Of course there are always losers in water transfers besides the buyers and sellers. At least in theory, economists have devised economic tools and incentives through which such losers can be compensated so that no one is made worse off.

It is not our intent here to reprise the many arguments criticizing the economic efficiency framework as applied to water resources. A number of scholars, including the editors of this book, have argued in the literature that water is not simply a fungible volumetric commodity—we maintain that water holds a multitude of values, and that the economic maximization of water captures only one of them—a form of value that is often overemphasized (Blatter and Ingram 2001). For instance, water is an ecological element that is always critical to biological species habitat. One specific example of species habitat broadly conceived is the fact that water is essential to the welfare of human communities. Any strategy to resolve global water problems must rely importantly on attaining greater efficiency. Yet, policies in the sole pursuit of water efficiency are bound to fail either because such policy proposals become bogged down in conflict at the conceptual stage or because efficiencies cannot be implemented in practice on account of the opposition among stakeholders that has been set in motion.

In the more than two decades during which the efficiency framework has dominated scholarly and practitioner discourse and action, considerable inroads have been made in a number of areas. Supporters of the application of efficiency principles to water resources decisions are able to proclaim correctly, "Since the 1980's when economists and policy analysts began to recognize that water markets could help allocate water, we have come a long way" (Anderson and Leal 2001, 93). Instead of pursuing new water supplies through the construction of dams and diversions, conserved water and water moved out of agriculture have become major contributors to meeting rising water needs. New water supplies from agriculture are transferred to urban users, often though leases or sales.

Economic incentives rather than regulations have been used to change polluter behavior. Some highly bureaucratic and wasteful municipal water utilities have adopted more economically efficient practices or been replaced by private water providers. At the same time, water problems have continued to worsen almost everywhere. Other considerations usually limit the pursuit of efficiency so that the results in the real world are very far from the ideals of economic efficiency analysis.

Consider the case of water in California that, from an efficiency standpoint, is badly misallocated. Water in the rapidly urbanizing South Coast area of Southern California is much more scarce than in the relatively water rich north of the state. Nearly two decades ago, economist and water expert Henry Vaux examined the exploding growth and water demands in Southern California and cast doubt on policy alternatives that relied heavily on new construction (Vaux 1988). The analysis by Vaux indicated that water users in the South Coast basin would be unwilling to pay the incremental or marginal costs of supplies made available through off-stream storage or other cross-delta facilities in the San Francisco Bay/Delta region. At the same time, taxpayers were unwilling to underwrite costs of construction with large subsidies.

Preferred alternatives identified by Vaux (1988, 363–373) included marginal cost pricing for water that would encourage substantial urban water conservation, and water markets to facilitate interregional transfers. In the years since that analysis, only some of this advice has been taken, results have been mixed, and sustainable water supplies for the South Coast of California are still highly uncertain. Reviewing the mounting water challenges to Metropolitan Los Angeles and its environs between 1990 and 2004, Steven Erie (2006) foresees mounting problems of all kinds, including water supply deficits, and places more faith in ingenuity of long established water agencies than the application of efficiency logic.

While cities on the South Coast have incrementally raised water rates and promoted water conservation, population growth and the urban water demands that follow from that growth continue to outpace the results from increased conservation. A statewide Water Use Efficiency Program was found to have fallen considerably short of its goals (Ingram and Lejano forthcoming). Municipal water utilities placed rate increases far down on their list of "best practices." Other strategies, including plumbing and fixture codes, public education and conservation appeals, are relied on more heavily than rate increases. Presently, more than half of California's water providers use a flat or declining block-rate structure, which rewards the highest water users with the lowest rates (Glennon 2005, 1883).

The Metropolitan Water District, a huge conglomerate of municipal water users in Southern California, has been successful in acquiring large quantities of water through sales and leases from irrigation districts (Anderson and Leal 2001, 93). Still, markets have yet to supply much more than 10 percent of the total supplies. Increases from that baseline depend on the availability of pumping capacity and conveyance through the California Bay/Delta region—an option that is highly controversial. The transactions that have taken place do not reflect the preferences of individual buyers and sellers as envisioned by the theories of efficiency. Fully 90 percent of the water purchasers are governments (municipal, state, and federal) motivated by many values besides efficiency.

It is much cheaper to prevent water pollution than to clean up previously contaminated waters. In the abstract, economic incentives are the preferred policy option to protect water quality. From an efficiency perspective, the adoption of "polluter pays" principles is an appropriate way to reduce the degradation of water quality. Referring again to California as an example, polluted runoff from storm events and other nonpoint and point sources is spoiling and degrading beaches and shorelines despite billions spent in waste treatment facilities and controlling industrial effluents. Eliminating pollution becomes a much more challenging task when sources of pollution are not large, but involve many individual polluters. Contamination from building sites, general construction, freeways, parking lots, fertilized lawns and the like is both hard to quantify and to assess charges to diffuse contributors. Moreover, as Kamieniecki and Below demonstrate in chapter 3, implementation of the polluter-pays principles falters when polluters and the beneficiaries of clean-up reside in places that are far apart.

Despite the commitment of governments and international institutions like the World Bank to water markets, there have been relatively few international, out-of-basin transfers (Gleick et al. 2002), especially transboundary transfers. There are emerging possibilities for water transfers from Austria to the rest of the European Union. Spain has under consideration a pipeline from the Rhone River in France to Barcelona. Israel is negotiating to buy water from Turkey, although Middle East conflicts have dimmed possibilities (Boulton and Sullivan 2000; Rudge 2001). Many of the impediments are the same as those affecting any large-scale water-basin transfer. There are other more fundamental disincentives as well. A nation's water tends to be thought of as part of its patrimony (Sandford 2003). The uproar caused by the North American Water and Power Alliance's proposal to ship water from Canada, Alaska, and the Pacific Northwest to the Desert Southwest is instructive. In that case, the United States Congress passed a moratorium on even considering the idea (Ingram 1990).

The efficiency framework suggests that private enterprises are more likely than public institutions to perform rationally. The policy prescription is privatization for addressing bureaucratic public water utilities that display inefficiency and corruption. Endorsements by the World Bank and other water reformers of privatizing water delivery services is relatively recent, but private ownership of water delivery systems has a long history. Up until the end of the nineteenth century, private water purveyors were dominant. Local governments began to take over water utilities in the 1880s because private owners tended to make initial investments that were too small, neglected maintenance, and failed to provide adequate service to poorer districts where profit margins were nonexistent or even negative (Gleick et al. 2002). Since the wave of water reforms beginning in the 1980s, private ownership of water utilities has increased markedly outside of the United States and Canada where adoption has been slow. Two French international firms dominate the private sector and have interests in water projects in more than 120 countries (Gleick et al. 2002, 24).

Private ownership follows different models, many of which are public-private partnerships. The experience to date suggests that whether public or private, utilities work best when a strong, accountable municipal government maintains oversight. As Robert Glennon (2005, 1896) states, "For privatization to be successful, governments must regulate water as a social good, ensuring access to all at a fair price." The case study of water utility privatization in Cochabamba, Bolivia by Madeline Baer in this volume illustrates that citizens expect to be able to participate in transparent and open decision-making processes in relation to their water service. Further, to be acceptable, utility decisions must have consequences that people perceive to be equitable.

Since the efficiency turn in water resources theory and practice, efficiency gains are not nearly sufficient to alleviate the serious global water problems that confront humankind. Markets work well only under conditions that are not as yet common throughout the world. Private investment is inadequate to respond to needs, even if there were less controversy about privatization. While important progress has been made in some areas, reform ideas adhering to efficiency formulations have also raised serious concerns and resistance. Efficiency is an important principle in policy formulation. But efficiency is no more than one metric to be considered within a broader effort to facilitate an equitable justice in environmental politics, and for this volume, in the principles which undergird allocation determinations about water from a global perspective.

Historical Roots of Equity

Efficiency has a history in Western political theory that goes back at least to Bentham and Smith in the eighteenth century. Equity has a considerably longer history. Equity is part of the tradition of common law and ideals arising from particular cases of water deeply attached to its social and environmental setting. The principles of equity are complex and contingent on circumstances, varied and nuanced, and cannot be fully understood until put back into the life cycle of living things. Consequently, there is no simple principle or set of principles, like those guiding efficiency, which can be set out as rules and universally applied in all places and circumstances. Instead, equity is a complex and protean idea. Like the concept of democracy, equity is not some objective state of being, but rather an ideal, vision, or aspiration that continues to challenge citizens to strive toward achieving it in greater depth, scope, and authenticity (Dryzek 1997). Justice is a broader concept than equity and embraces fairness, dignity, respect for mutual rights and obligations, and ensuring that institutional arrangements—for managing water, as for other necessities of life—nurture the full development of human capacity. Equity is a necessary condition for a just society. The search for equity requires deliberation, discourse, and inclusive argument such as we provide in this book.

Among the arguments bolstering equity in relation to water is its roots in antiquity and long-established practice. Classic philosophers concerned with establishing the conditions for a just society recognized that water is an important part of its health, defense, and beauty. Several passages in Plato's *Laws* argue that of all the resources and necessities of life, water is the most basic to human well-being. Water is vulnerable to "doctoring, diverting or interception of supply," and water "must always be subject to public regulation." Much the same association of water to the just society is made in Aristotle's *Politics,* which observes that public management of water "is a matter which ought not to be considered lightly." According to Aristotle, sufficient water supplies for life need to be guaranteed to all householders, including slaves, before allocation to other economic activities like agriculture (Ingram, Scaff, and Silko 1986). These principles were widely appreciated throughout the ancient Mediterranean world. The Hebrews understood the importance of water to hygiene and personal health, recognized that the availability of water influenced the abundance of crops, and were admonished by the Old Testament to honor their sacred covenant with God both to ensure this continued abundance, and to serve as just stewards of the resource (e.g., Numbers 19–20; Deuteronomy 8:7, 8:15, and 29:19). In essence, the declaration of the United Nations making access to water a human right has very old roots.

There has been much written about the importance of the aqueduct system to the Roman Empire and the exceptional efforts made by the Romans to assure civilization to its far-flung outposts by providing adequate supplies of high quality water. Histories of the oldest irrigation societies on the planet identify fairness as the critical criterion to explain both the adoption and enforcement of water allocation regimes. Maass and Anderson (1978, 395) found during their study of water allocation that evolved in six irrigation societies in Spain and the American West that fairness carried considerably greater weight than efficiency in the design of institutional arrangements. Water apportionment policies reflected concerns with popular control, distributive shares, economic growth and farmers' conception of fairness (Maass and Anderson 1978, 395). Keeping the peace and maintaining trust among farmers was essential in drought-prone climates where water shortages requiring rationing were common. If equity concerns were paramount for these relatively simple societies made up primarily by farmers, how much more important must equity be to water governance in contemporary, more complex societies?

Fair Consideration of Multiple Values of Water

Equity dictates recognizing value differences and treating them evenhandedly. Among the utilitarian values that must be considered is efficiency, but no particular privilege need be attached to this consideration. As

Sheldon Kamieniecki and Amy Below note, utilitarianism itself requires the consideration of diverse equities. There are ends-based, or consequentialist standards that dictate actions which should be judged simply on results. According to these principles, the just action is that which leads to the greatest happiness. Some utilitarian thinkers believe that it is not enough for actions to improve the welfare of the greatest number. Additional alternatives need to be considered so that the best possible consequences will be reflected in choice.

Of course, what is judged to be a good consequence to some may be disputed by others. Further, there is a specific kind of utilitarianism that focuses on motives as well as outcomes. This framework may lead to contradictory conclusions about the equity of choices. According to Kamieniecki and Below, procedural utilitarianism requires some equality in access to decision making so that there is some chance that each person's happiness will be considered. There may be significant gaps between procedures and results. In chapter 4, Margaret Wilder argues that even when procedural utilitarianism (which she terms "political equity") is served reasonably well, other values need to be taken into account including the access to and affordability of water to the poor. Wilder finds that what appear to be democratic procedures instituted by water reforms in Mexico have led to discord and inharmonious equity that separate poor communities even further from productive resources than previously was the case.

For some people in some cases, the values associated with water are intrinsic, not utilitarian. Equity requires that these values be treated even-handedly, even though equity does not allow one set of values in the abstract to trump all others. Water flows through natural and human communities in such close association that abstracting it from its setting and rationalizing it by assigning a quantitative value is to do irremediable damage. In this reading of intrinsic value, humans, other living things, and water are inseparable. Ecosystem sustainability requires water for life for both human beings and animals in their natural habitat (Arrojo-Aguda 2006).

Symbolic, religious, and lifestyle meanings of water are among the diverse values that are poorly captured in any kind of utilitarian calculus. Yet they must be included in deliberation for equity to be served. Consider the Hindu association with the Ganges and the Christian ritual of baptism. Spectacles like Old Faithful in Yellowstone National Park or Victoria Falls are the patrimony of humanity. The native peoples, the Northern Apache, identify and bind their culture to place, and ancestral myths are believed to have occurred at particular locations. The invocation of a particular place name, such as where a brook bends around a large rock, may signal a moral lesson about flexibility (Basso 1996). To lose or alter physical place, including water, is to lose culture and the values that make the community distinctive.

Water has increasingly assumed a key position in defining healthy lifestyle whether in sports such as surfing or in designer label water bottles. Of course, symbolic values are socially constructed. Long established precedent weighs heavily in considering legitimacy of claims because over time a particular symbolic value of water becomes deeply embedded. Robert Sandford (2003) chooses the canoe as the appropriate object that binds together Canadians to their water-rich home. He writes:

> Of all the symbols that have survived from the period before European Contact in what is now Canada, the canoe is both the most unique and the most enduring. It has been said that if there is an intergenerational symbol of the sense of place shared by all the peoples and cultures who have experienced this land, it is the canoe. More even than the maple leaf, the canoe means Canada. The canoe is a symbol of the exploration and discovery that defined this nation. It is a symbol of our deep connection to our waters and the harmony that is possible in our relationship with nature. (Sandford 2003, 55)

Newer lifestyle symbols may carry less legitimacy in terms of serving equity than those instilled over long periods of time. Ismael Vaccaro expresses a distinct reservation about the postmaterialist values that fuel the movement of waterscapes in the Spanish Pyrenees from their historic uses in agriculture to their role in creating leisure spaces for tourists. Similarly, Robert Glennon (2005), who embraces much of the efficiency framework for making water decisions, distinguishes what he sees as the more legitimate claim of Coors beer to large quantities of "Rocky Mountain Spring" water from that of Nestle Waters North America intending to bottle approximately the same amount from the same source with the same refreshing symbol. The difference is a matter of past uses that should be grandfathered in versus proposed new or newer uses (Glennon 2005, 1897).

Past promises lend legitimacy to equity claims. Unlike efficiency, equity is grounded in the context of time, taking into account the historical sacrifices and past promises as well as hopes and dreams of a better future. Efficiency takes no account of past investments, treating them as sunk costs. In contrast, equity attends to covenants or other legal forms such as prior appropriation, or first in time, first in right. In order to secure investment for the development of the American West, state governments promised that diverters of water from water courses would not be left high and dry when another diverter moved in upstream at some later time. Without such guarantees, agriculturists would not have made investments in water distribution infrastructure.

The principle of equity suggests that past promises must be considered, even if they are outweighed by needs to provide equity to existing deserving but underserved populations. Similarly, opportunities for development exist for a region or people if future access to water is preserved. Equity dictates that present day decisions not unduly burden the scope of future human choices. In chapter 6, Paul Hirt considers the conflicting time-related equities of Native Americans, hydropower generation, irrigators, fishermen, endangered species, and other actors in the Columbia River Basin. Seeking for the antiquity or headwaters of equity claims in treaties and litigation resolves very little.

The rights of future generations in relation to those of the present generation are clearly equity matters. The efficiency argument is tilted toward weighing most heavily the economic interests of the present day. The argument is that if water is used efficiently, human welfare improves and future generations inherit higher levels of human welfare on which their own generations can build. While this argument has merit, especially in terms of theory, the many case studies in this volume dealing with the inheritance of past policies slighting equity suggests otherwise (see particularly chapters 5 and 6). Legacies of unfair treatment leave distrust, bitterness, and disinclination to cooperation that undercuts human capacity to deal with present and future complex water problems. Also, water has emotional and identity attachments not typical of other common resources like coal. Only some of its services can be provided to future generations through technological advances or other resource substitutions.

Intergenerational equity may be served by providing citizens with long term scenarios and models that project into the future trends such as climate change. Models of likely future withdrawals from river basins measured against possible reductions in supply may help refocus citizens toward longer term consequences of present day decisions. Yet, the case study from Brazil of Maria Carmen Lemos, which examines how watershed governance councils use such technical information, suggests that it may also serve to perpetuate the dominance of technical elites over water resources decision making.

Community Equity in Water

Unlike efficiency, equity applies not just to individuals but to groups of people and to communities in particular places. Moreover, equity may often demand that individuals look beyond their own welfare, recognizing that the good of the community may well be different from personal welfare. The community association with water is sufficiently prevalent in many communities to merit status as a distinct equity concern. F. Lee Brown and Helen Ingram (1987) examined the cultural and identity association of communities of rural Hispanics in the Upper Rio Grande Basin and Native Americans in Southern Arizona. They found that in the continued proximity of these communities to the natural environment, and in the insecurity of these cultures, the presence of water was absolutely crucial to their community well-being (p. 29). The experience of building and maintaining the historic *acequia* system through which water was shared by community members was transformative in building a sense of community among the residents of northern New Mexico.

An outgrowth of the merging of Spanish and Native American custom, the *acequia* system originated as associations of all persons served by the ditches, and required all members to contribute labor for maintenance and fair operation of gates so that even in times of shortage all could be served equitably. Far more than just a means of water distribution, the *acequia* was the fundamental organizing unit for rural peoples in northern New Mexico. It united people across lines of race and class (Brown and Ingram 1987, 49). The extent to which community members built their identity around water became the theme of the John Nichols novel (1974) and later movie, *The Milagro Beanfield War*. This fictionalized account, drawn partially from real events, portrayed attacks on local control over water brought by state and federal governments to be threats to culture and community (Brown and Ingram 1987, 61).

For the Tohono O'Odham in Southern Arizona, the ongoing loss of groundwater to the city of Tucson was symptomatic of a long series of sacrifices including loss of land, loss of ability to move freely across the international border that isolated some of their members, and loss of the political autonomy of individual bands imposed by the federal government's wish for administrative simplicity. In the legal and legislative battles to regain groundwater rights, what was at stake seemed much less related to exercising community control than to securing the long-term survival of their community in the face of continual siege (Brown and Ingram 1987, 104–177). The idea that security for communities is a value at stake in water controversies is supported by other studies. For example, an opinion survey of community leaders in communities on the receiving end of proposed rural-to-urban water transfers (El Paso, Texas, and Tucson, Arizona) and their counterparts in rural areas of origin were in agreement that when water rights were lost, the community sacrificed opportunities in the future that could not be compensated adequately by money transfers (Oggins and Ingram 1989).

Clay Arnold moves the discussion of community equity in water resources considerably beyond previous understandings. Rather than referring to cultural and symbolic associations between water and community, Arnold analyzes how a sense of community is generated through water. In discussing the collective relationship of the residents of the San Luis Valley in Colorado to their water, not only is their association tied to the historic and ongoing *acequia* system, but also to continuing struggles to assimilate successive waves of immigrants and new institutions without losing the "sense of attachment and mutual obligation by which they recognize themselves as members of a community" (Arnold, chap. 2). Arnold names "moral economy" as the construct to describe the difficult reconciliations that are made between competing values related to water in the community.

A far richer concept than "social capital," the community's "moral economy" relates to its capacity to act collectively and successfully on the most challenging water conflicts. The "moral economy" of the San Luis Valley, particularly the networks of associations built up over a succession of struggles, and the active sense of shared fate, becomes easier to mobilize on each successive challenge. Successive mobilization strengthened rather than weakened community. This circumstance allowed the San Luis Valley to handle skillfully the legal and political challenges of powerful and astute adversaries. In addition to honing decision-making skills as part of the "moral economy" of the community, the habit evolves of thinking about water issues in complex rather than one-dimensional terms. Arnold argues that complex equity opposes reductionism and furthers the justice of nondominance. The hard-earned "moral economy" achieved by members of the community simply rejected as inappropriate any economic of efficiency solution. The community in the San Luis Valley favors policy choices that reflect both inclusion of, and balance among, diverse values.

Inequitable Water Development and Redressing Distributional Inequity

Water has long been used as an engine of regional economic development. While construction of new dams and reservoirs has slowed in the United States and Canada, such construction continues in Turkey, China, Brazil, Malaysia, India, and elsewhere. Moreover, there is still a lot of other water resources construction ongoing including flood control structures, wastewater treatment plants, water-based recreation facilities, and environmental restoration of wetlands. The World Bank and other donors are pursing urban water supply and sanitation projects in many developing countries, often in the name of poverty alleviation (Bhatia and Bhatia 2006).

While pursued in the name of consequentialist utilitarianism, only a minority of water development projects could survive a formal test of efficiency that included all direct and indirect costs (Bhatia and Bhatia 2006, 217). Yet, water development appears to go hand in hand with prosperity so often that belief persists in the Midas touch of water. Examples are legion: the enormous boost given to the Pacific Northwest through hydropower development; the head start gained by the city of Los Angeles in laying claim to abundant, cheap water from the Owens Valley to fuel population and economic growth; the economic development and uplifting from poverty of the mid-South's Tennessee Valley through the federal Tennessee Valley Authority; and, the dependence of New Orleans for an economic future on dam and levee reconstruction after Hurricane Katrina.

Uplands watersheds tend to be treated inequitably in water resources development. Historically, mountain peoples have been displaced by dams backing water up into mountain valleys and flooding out residents. Displacement today is different but no less disrupting, as high-income vacationers attracted to water-based recreation crowd out natives with longstanding cultural roots in the mountains. Chapter 8, by Ismael Vacarro, about modernizing mountain water in the Pyrenees in Northern Spain, tells this story. Many environmental services such as storing water in snow and icepacks, preserving wetlands that buffer downstream floods, water quality protection, and silt control are provided for free or for low cost to downstream residents by upland regions. Global climate change that is disproportionately threatening higher elevations is drawing long needed attention to upland water problems (Rosenberg Forum 2006).

Many water developments fail to satisfy the basic distributional equity and environmental justice tenet that no groups, particularly the disadvantaged, should be made worse off in absolute or relative terms because of water policies. It is unfortunate that the treasure which cities, states, and nations continue to sink into water projects, if invested elsewhere, could provide higher economic efficiencies. The more profound tragedy is the inequity in the distribution of benefits and costs of many water development projects. Several chapters, especially 4, 5, and 6, provide specific, historical analysis of how injustice came about. Developments including canneries as well as dams on the biologically rich rivers in the Pacific Northwest broke treaty promises to Native Americans and deprived them of their livelihoods and way of life (chapter 6). The risks to farmers in Mexican borderlands, especially those most economically vulnerable, grew as a result of historic water development projects that stretched water supplies beyond ecological sustainability (see the Mumme and Wilder chapters, 4 and 5).

The construction by the United States of the Welton-Mohawk irrigation project was done with disregard of the impact of the salinity runoff from the irrigation project on water quality delivered to the Mexicali valley. This disregard of equity on the part of the United States continues with the lining of the All-American Canal, which deprives Mexico of waters that probably belong to Mexico under international law (chapter 5). As the authors of these cases demonstrate, appropriately considering distributional equity in water development decisions would have resulted in avoiding debilitating conflicts and lingering distrust among peoples who must cooperate if water problems are to be resolved.

Flagrant disrespect for equity might appear to be less of a problem as large-scale water development projects have been scaled back. Yet, equity problems persist in the handling of the water quality problem, and under new laws instituting water reforms. The application of the "user pays" principles to water quality problems presents equity problems as chapter 3, bySheldon Kamieneicki and Amy Below, shows. Chapter 4 is a rude awakening for advocates of market-based water reforms. Examining the consequences of water resources reforms in several areas in Mexico, Margaret Wilder finds that poor farmers have become more vulnerable to competition from larger and more efficient farmers in Mexico and elsewhere as a result of free trade. Further, the ability of poor, small farmers to retain collectively rights to water resources are undercut by reforms allowing more liberal sales and exchanges of water collectively owned by poor communities (*ejidarios*).

The reforms would seem to have provided better political forums for poor farmers to express their interests, but the problems of poverty rob them of the resources necessary for effective political participation. Poor Mexican farmers are even further alienated than previously from the very resources they need to bring themselves out of poverty.

In fairness, it must be observed that many of the inequities suffered in the water developments considered in this book have come at the hands of political rather than market driven processes. Inequities can be traced to asymmetries and imbalances of political as well as economic power. Chapter 8 reminds us that water policies are related to the larger repertoires and natural resources policies of the state. States have often exercised their power on behalf of narrow, special interests. Politically subsidized water to farmers, urban residents, and hydropower users threaten sustainability as well as equity. Water development projects generally have not served the interests of indigenous peoples. Further, politically driven water projects have encroached on the water rights reserved to native peoples, and, at least in the United States, the federal government has been a poor steward of trust obligations to Native American tribes.

Political power, or the ability to marshal support in authoritative venues such as the branches and levels of government and media, is often decisive. Efficiency and equity may be important rationales employed in the process of building and wielding political power. But there are many cases in which political power serves neither of these values, yet still prevails. The state of California, for example, has long used more than its fair share of Colorado River water, and has succeeded in getting guarantees that it will continue to get its full legal entitlement even in drought years when all other Colorado River basin states have to cut back their uses. The strength of the California voting block in Congress, and the expertise the state has been able to command in administrative settings and the courts have trumped competing claims from other contenders for water. As a consequence, California was the first state to fully develop the Colorado River with huge dams linked to aqueducts moving water at federal government expense to supply cities and farms located hundreds of miles from the main stem of the river. The association of the economic success of Southern California with water development has served as a vivid lesson to other boosters of development in arid areas that ample supplies of good quality, cheap water are necessary for a brilliant future.

Compelling equity arguments can attract public support and therefore political power in some circumstances and venues (see chapter 7, on Cochabamba, by Madeline Baer). Indigenous peoples have sometimes been successful in courts and congresses in asserting equity claims to waters expropriated by white settlers. The argument of native peoples becomes more powerful politically when they are socially constructed as deserving in the eyes of public opinion. The romantic images of noble Native Americans created in literature and cinema have done much to reconstruct peoples once thought to be savages in need of total assimilation. In a similar social construction process, the equity rights of small farmers, fishermen, and others have become more positive through media portrayals, even though these groups lack efficiency arguments or economic power.

While politically driven decision-making processes often involve inequities, political processes, including the courts, do provide opportunities for mobilization and protests on equity grounds. Many government project appraisals around the world now include benefit-cost analysis, administrative review procedures, and environmental assessments. Environmental Impact Statement processes in the United States provide venues and forums to raise equity issues. Environmental justice offices in federal and some state agencies, and the Bureau of Indian Affairs, can flag equity problems for further debate. While fewer big projects would have been built without government involvement, private projects typically have not provided the same kind of structured opportunities to raise equity concerns.

Institutions charged with protecting the interests of the poor and disadvantaged groups need to balance the opportunities as well as harms to their clientele, and assure that processes of review are open to the disadvantaged to express meaningfully their own preferences. While equity does value diversity and variety, it does not require that indigenous peoples retain ancient practices in environments where such practices no longer lead to happiness and productiveness. Water scholars are well aware that much has been learned from "backward" peoples about environmentally sensitive and technologically appropriate water practices (Ostrom 1990). If for no other reason than to provide viable alternatives to dominant cultural practices, access for indigenous peoples to water and to water based livelihoods needs to be reevaluated seriously.

Alternative schemes for water development and water reform involving markets and privatization present equity problems in areas of water origin. Poverty alleviation on water investments may well require postponement or even shelving efficiency notions of full cost recovery. In addition, lifeline water rates that levy only token charges to the poor for urban water delivery serve equity. Also serving equity, and to offset low rates to poor users, is the setting of high water rates to the well-off, serving as a form of subsidy of lifeline water rates (Bhatia and Bhatia 2006, 216). Government regulation of markets through conditioning marketing permits can mitigate some of the inequitable effects of inter-basin water transfers. Many states have passed legislation protecting areas of water origin by limiting the amounts of water that can be transferred out, imposing permit or other requirements on diversion and return flow, requiring the consulting of communities affected, and insuring that compensations are both provided and acceptable.

Equity across Boundaries

There are many kinds of physical, political, and social boundaries. Whether and how those boundaries are drawn have enormous implications for equity in the allocation of water resources. River basins and watersheds have physical presence and systemic properties. Thus, what happens in the headwaters of a river system has clear implications for downstream flooding, water quality, and the environmental health of riverine species and estuaries. Despite this connection of physical boundaries and their broad implications, political boundaries frequently create fragmented management and introduce serious problems of transboundary governance. For many years the systemic aspects of river basins and watersheds, in terms of the transboundary governance structures and their broad implications, were ignored. Uncoordinated actions resulted in severe environmental damages. Even when there are adequate arrangements for watershed management within states, there continue to be equity problems in upstream-downstream distribution of benefits and costs, as chapter 3 demonstrates. Further, decentralization of power to basins and watersheds does not assure that forums at lower levels will sufficiently reflect equity values (see chapter 4 for a clear example).

Many rivers cross international boundaries. There is a large literature on transboundary water, including another book in this series (Blatter and Ingram 2001). Chapters 5 and 6 in this volume contribute importantly to the understanding of equity in treatment of transboundary waters. Nation-states are reluctant to relinquish sovereignty or to honor international norms of equitable apportionment or utilization. The experience of the United States with its neighbors to the North and South suggests that great powers are more likely to honor equity principles when the asymmetry between the wealth of nations is not large. Mexico is twice cursed with a neighbor that has greater resources in international relations as well as citizens who are much less poor. As Stephen Mumme explains, the United States once ascribed to the Harmon Doctrine that essentially allowed it as the upstream party to do whatever it wanted to develop the Rio Grande and the Colorado Rivers. Both because the power differences are not so great between the United States and Canada, and because hydroelectric power revenues as well as control over water are involved, there are more resources with which to offset injuries. Paul Hirt sees the relations between the United States with Canada in the Pacific Northwest to reflect a far greater concern with equity.

Equity and environmental quality may be aided rather than hurt under some circumstances. Paul Hirt, in chapter 6, on relations between the United States and Canada, argues that the location of the Fraser River entirely in Canada, as opposed to the Columbia River, which is shared with the United States, facilitated the maintenance of salmon on the Fraser. Initially alone by the United States, and then later with the assent of Canada, the entire Columbia River was developed for hydropower, essentially dooming the huge salmon runs that once made up between 30 and 80 percent of the diet of indigenous peoples living in the Columbia River basin. Canada learned from the experience on the Columbia River. The existence of the international border insulated it from pressures to expand hydro-electric development to the Fraser. However, Hirt observes that a battle continues between the United States and Canada on the rights to catch what are essentially Fraser River fish in international waters.

While domestic and international law often uses equitable utilization and apportionment as operative criteria in treatment of shared waters, negotiators of transboundary compacts usually recognize only a narrow range of economic values as legitimate. Professors Hirt and Mumme, in chapters 5 and 6, note some encouraging trends that recognize equity, public participation, and sustainability as factors in the management of bi-national water resources. Subnational regions that encompass several states or provinces also exhibit some positive movement. There are examples of regional water management that encompass whole river systems. The Murray Darling Basin in Australia is an example of a strong basin institution dedicated to ecological restoration. Similarly, the California Bay/Delta Authority and the Everglades in Florida are examples of multiparty regional ecological restoration arrangements involving not only states and localities but also the federal government and nongovernmental entities. The success of these efforts depends strongly on the ability of the participants to engage in a unifying basin-wide vision that transcends political jurisdictions and embraces equity. An essential problem is that these examples of successful transboundary governance are the exception rather than the rule.

Recognition of social and political boundaries sometimes greatly advantages place-based minority populations asserting equity interests. Tribal sovereignty over natural resources including water has been enormously important to the ability of Native Americans to preserve Indian reserved water rights. While it has often been difficult for native peoples to turn their "paper" water gained through court decrees into usable water, the territorial integrity protected by reservation boundaries has been critical to what gains have been made. As chapter 6 attests, indigenous peoples have not been able to hold on to their water related rights, including fishing, in face of efficiency and development pressures despite having official recognition, and dedicated homelands. There are signs, however, that indigenous rights in shared water resources are being afforded more protections, although as Paul Hirt observes, recompense for past injustice is difficult in the present era of scarce water and limited financial resources.

Equity, Participation, and Process

Equity requires fair, open, and transparent decision-making processes in which all individuals and groups affected by water decisions have an opportunity to participate. David Feldman has written that "no ethical approach to the management of water resources should be adopted that categorically excludes any constituency or alternative approach to management out-of-hand. This means that any approach to management should emphasize process as much as substance—providing the widest possible debate and deliberation" (Feldman 1991). Participation must be meaningful so that choices made in participatory forums actually matter and are taken with utmost seriousness. Participatory rituals are not to be performed simply for cosmetic purposes. Participants need to have the necessary resources with which to participate in terms of information and time so that they can afford to engage fully. Participatory processes ought not to be dominated by professional and economic elites. Nor should such processes be subject to procedural manipulation that distorts the equity issues in question. In one way or another, almost all of the chapters in this volume link equity with process and procedural fairness and reject the false notion that good results can excuse unfair processes.

In explaining what he calls the "moral economy" of water practices in the San Luis Valley, Clay Arnold documents why the multiple values that are associated with water must be reconciled in conflicts over the scarce water resources that have historically plagued this Colorado community. The process of reconciliation is critical. Early on, valley residents constructed a forum, the Rio Grande Conservancy District, through which collective decisions could be made that both reflected the community view and stood for community interests in relation to state and federal agencies and to private interests as well. This institutionalized voice of the valley was supplemented by a number of other grassroots organizations that drew together valley residents in face of outside threats to water security. San Luis Valley residents objected to both water markets and state-wide referendums as inadequate forums through which to make community water decisions, the first because it was too narrow in its representation of only the economic values at stake; and the second because it was too broad and involved a slightly engaged state-wide public vulnerable to manipulation.

From the perspective of the "moral economy" of the San Luis Valley, appropriate decision forums ought to represent fairly the way of life of the valley, and ought not to represent overly the economic interests that would benefit from water transfers out of the valley to front-range cities. Equitable arenas reflect the complexity of values associated with water, not overly simplified efficiency claims. When proponents of water transfers tried to manipulate public opinion by claiming false benefits of water transfer schemes to the public schools in the San Luis Valley, mobilized citizens responded by unmasking the strategy as mainly benefiting out-of-state speculators. Clay Arnold in chapter 2 observes "Decisions and decision-making processes that fail to regard complex social goods as complex commit a kind of injustice; they conceive the good as something other and less than what it is for many of the individuals and communities directly affected. More precisely, they conceive the good incompletely and therefore unfairly."

Even a fairly narrow notion of utilitarian standards related to water that marginalizes intrinsic values recognizes procedural equity or equality, although not very satisfactorily. Sheldon Kamieniecki and Amy Below write: "The utilitarian perception of procedural equality infers that everyone is afforded equal opportunity to affect policy decisions as each person's happiness is weighted . . . there is often little actual utilitarian value in procedural utilitarianism. Everyone's opportunity to affect policy outcome is not equally weighted. Some are provided more opportunities while others are victim to others' decisions." As Hans Morgenthau is quoted in chapter 3, "the test of a morally good action is the degree to which it is capable of treating others not as means to the actor's ends but as ends in themselves" (Morgenthau 1945, 14). By the procedural utilitarian standard, equity has not been present in much of the history of water resources decision making in many parts of the globe.

It is totally unwarranted to expect that somehow markets will deliver procedural equity. Equity assumes that affected interests will participate in decisions. One of the most compelling arguments against efficiency is that it maximizes the power in decision making of those with financial and intellectual resources. Clearly, market participation favors those with money and can afford to wait until they get the right price before buying or selling. Further, efficiency in government programs is often driven by benefit-cost analysis that privileges the exercise of experts including engineers and economists. Economic returns not community value of water or other equity issues drive consideration of efficiency.

Contemporary water reformers, supported by such institutions as the World Bank, have looked to the involvement of economic incentives and the formation of watershed governance for insurance of procedural equity. Chapter 7 carefully documents how failure to follow open, fair, and transparent procedures in the privatization of a municipal water utility and in setting higher water rates led to protests and violence in the streets. In contrast, in chapter 5, Steve Mumme concludes that the more participatory forums, which have been created by recent international agreements between the United States and Mexico, have given border residents more control over their lives and resulted in greater consideration of sustainability in water decisions.

Other chapters concentrate on more subtle breaches of procedural equity in the course of implementing water reforms. In chapter 4, Margaret Wilder contributes importantly to our understanding of the requisites of fairness in process. She examines the widely praised new water laws in the border state of Sonora, Mexico that facilitate grassroots governance of watersheds. What Wilder calls "political equity" afforded by watershed and river-basin planning institutions was not meaningful because poor farmers lacked the resources necessary to make their participation count. Further, decentralizing decision making over water resources at the regional and watershed levels is not a significant improvement in terms of equity if the economic resources necessary to redress distributional inequities are not reallocated also from the central to local governance.

Wilder judges that the positive effects in terms of procedural and participatory equity are more than offset by the drastic decline in economic welfare of poor farmers as they struggle to remain in active farm production under a nexus of economic pressures including trade liberalization, lack of subsidies and state credit, and a growing indebtedness that, for many, is insurmountable. Although they have some of the requisite tools to succeed—including irrigation, access to technological sophistication and knowledge, the opportunity for integration into commercial export economies, and location just miles away from Mexico's largest trading partner, the United States—these poor, small farmers are unable to compete in the market-driven international trade dominated by those with superior resources. Without an economic basis for survival, procedural improvements really are not meaningful.

The dominance of technocratic elites over what is considered legitimate information for water resource decision making is also a contemporary threat to procedural and participatory equity. Watershed governance is supposed to counterbalance the long-term dominance of nationwide-based water resources bureaucracies that have controlled traditional water development. Better informed stakeholders should be able to make wiser decisions. However, if knowledge is controlled by a few actors who mainly use it to bolster their own decisions, technical knowledge can then be used to exacerbate power imbalances between those with access to knowledge and those without.

In chapter 9 on watershed governance in Brazil, Maria Carmen Lemos raises serious questions about the procedural and participatory equity actually realized through watershed governance reforms. She reviews new evidence demonstrating that despite more participation, the inclusion of nonelites—such as small farmers, rural workers, and rain-fed farmers—has been thwarted both in terms of representation (they are less represented) and influence (they exert less influence during the allocation meetings). She also finds that in the cases she studied of watershed reform in Brazil, the handling of technical information led to "elite capture" of decision-making processes, which in turn affected broader issues of equity and justice in water management. Despite the effort from local *técnicos* to improve communication and the availability of techno-scientific information, Lemos found evidence that a substantial number of stakeholders find technical information neither available nor accessible. Moreover, there is a widespread perception of *técnicos* as the most powerful actors in the water management process.

Interspecies Equity and Sustainability of Water Resources

Notions of equity and environmental justice have generally been related to people and communities, not to the plant and animal world. It is not our intention here to engage in the fascinating debate between deep ecologists, animal rights activists, and others about whether trees and bears have rights. Certainly the Endangered Species Act suggests that when the survival of a living species is at stake, the presumption is in its favor rather than on the side of human desires.

Neither the ethnocentrists nor the ecocentrists have much to offer to understand water and equity. Water is inextricably bound up in all life. Astronomers look first for water as evidence that a distant planet could support life. We are in water before we are born. For Robert Sandford (2003, 12) who survived being swept under a glacier before reemerging in the North Saskatchewan River, the birthing process happened twice. For most of the rest of us, waterscapes ranging from pounding oceans to bubbling streams are sources of joy and renewal. The European Declaration for a New Water Culture, signed by one hundred scientists from different European Union countries in 2005, was correct in declaring water for life for both human beings and animals in their natural habitat as one of four ethical categories of universal rights (Arrojo-Aguda 2006).

While it sometimes appears that environmental interests and equity for human communities are in conflict, a close historic reading of how the conflict developed almost invariably shows prior disregard for equity. Clay Arnold's history of water controversy in the San Luis Valley in chapter 2 is illustrative. Attempts to transfer both groundwater and water rights out of the San Luis Valley to serve the demands of cities along the Front Range of the Rockies threatened irrigation-based agriculture, and would have changed the Valley's culture and way of life. The proposed transfers would have destabilized the Great Sand Dunes in the national park. The proposed transfers were pursued by processes designed to short circuit public participation. Happily, water in the San Luis Valley is today recognized as a complex social good. The water and rights at the heart of the controversy are relatively safe in the National Park system.

The poor fate of salmon in the Pacific Northwest is shown to be closely tied to the broken promises made to Native Americans. At the end of chapter 6, Paul Hirt refers to the contemporary discourse of regret in which many credible voices question whether the social and environmental sacrifices made for northwest hydroelectric power development was worthwhile overall. The relatively unspoiled environment of the Fraser River that was never dammed is a reminder of the path not taken.

Chapter 5 tells the same story of the shared fate of disadvantaged people and the environment along the United States–Mexico border. In its preoccupation to put to use economically as much water as possible from the Colorado River, the United States has overallocated water rights so that in dry years endangered species like the silvery minnow in the Upper Rio Grande River is driven to the brink of extinction. Plans to increase water use efficiency of Colorado River water in the Imperial Valley by lining the All-American Canal has both environmental and equity consequences to farmers in Mexico that have not been open to public discourse and consideration.

Unfortunately, humans, nature, and fairness are knotted together. This knotting together is a fact of life on this planet. The failure to recognize that fact has created the unfortunate inheritance chronicled in this volume of a past not sufficiently conscious of equity issues. Among the most tragic of contemporary conflicts are those between the equities of the interests left behind by the era of large-scale water development. Native peoples now wishing to develop water-based rights and activities find themselves at loggerheads with environmentalists trying to protect the few wild places that have been left undisturbed. The reason, unfortunately, is that the few wild places are often located on tribal lands passed over in the previous era of water development. Balancing equities to species and to disadvantaged communities necessarily involves compromise, including just compensation to parties forced to relinquish equity claims for the greater public welfare.

What kinds of burdens and regulations should be placed on the water uses in areas that prevail in equity contests? Ought the broader society whose actions in the past resulted in the necessity of making such difficult choices in the present share part of compensatory burdens? What is "just compensation" for losses of intrinsic values? Does the efficiency principle of "beneficiaries pay" serve equity in difficult balancing situations? The processes through which conservation is implemented in communities have a great deal to do with their effectiveness.

Equity dictates that that just compensation will be provided for those who make sacrifices for the good of the whole including other species. The important distinction between compensation provided in efficiency-driven decisions and those where equity comes first is that those who receive compensation are the judges of sufficiency. It may be, for example, that losers in a water decision may be placated by promises of benefits in the future. The equity proof here is not whether humans are better off over the long term, but whether those who make sacrifices believe their sacrifices will ultimately be meaningful in terms of long-term sustainability of water resources.

Equity and Water Ethics

The efficiency framework assumes that the water user will serve his or her self-interest in relation to water. People are expected to respect the economic value of water. Waste is counterproductive but not necessarily unethical. The editors of this volume join with many others, including the signatories to the European Declaration for a New Water Culture, who see water ethics as much more demanding than this. In the words of Pedro Arrojo-Agudo, "it means recovering the holistic perspective embodied in the Aristotelian concept of 'economy' and going beyond narrow mercantile approaches which dominate the reigning model of globalization" (Arrojo-Aguda 2006, p. 24).

Equity as discussed here goes beyond self-interest to include concern with serving broader community values, the effects on the poor and disadvantaged, and respect for equitable processes, and broader issues of justice, such as dignity and the fulfillment of human capacity. Equity considerations suggest some fairly obvious ethical principles that have already been discussed, such as, that every human being should have access to sufficient water to maintain life and health. Community interests and respect for value diversity in water reflects the broader equity principle of solidarity, that is recognizing each person or group has a right to participate in open and inclusive decision-making processes so that their values and visions can be considered fairly (Priscoli et al. 2004).

Distributional equity obligates the husbanding of water resources. One ought not to use all of the water one can afford, but instead use only as much water as one needs and can ethically justify in the particular contexts of scarcity. Water needs to be used so that it is not unnecessarily degraded by contaminants. Ethics require that one have some knowledge and concern that sources of water supply and means of disposal of water once used are in accord with the underlying values associated with the resource. Concern with intergenerational equity suggests an ethic of stewardship. Each person in each generation should treat water resources so that they are passed along to future generations undiminished—bearing in mind that this is precisely what we would wish prior generations to do for us.

A water ethic requires one to participate in such a way that discourses and actions reflect more than self-interest. A sincere search for common interests requires a spirit of cooperation in seeking to identify broader societal interests. Because water is integral to so many human activities, the smooth running of water systems depends on many interlocking networks of people. Water management is vulnerable to disruption, therefore, at many points by people who refuse to cooperate. However efficient the policy design, policies are unlikely to be implemented without cooperation (Priscoli et al. 2004).

The European Declaration for a New Water Culture provides four broad categories for ethical action. In addition to the "water for life," both human and nonhuman, the declaration elevates "water citizenry" so that water becomes an instrument for maintaining not just water supply and sanitation for health, but also for well-being, social cohesion, social capital, and capacity-building. The Declaration further recognizes the importance of "water business" for economic growth, but this is a third level of priority and it is unethical to allow these business concerns to interfere with water for life. Finally, the Declaration takes a firm ethical stance against the "water-crime" nexus that has led to destructive withdrawal practices, toxic spills, and other actions that threaten the globe's precious and irreplaceable water resources (Arrojo-Agudo 2006).

The Contributions of This Book

While equity is a deeply embedded concept in water resources, it has often been treated in water research as a kind of residual category of concern that is taken into account when all else is equal. As a consequence, equity principles are not as well articulated and accepted as they should be. Further, there is insufficient research on the nuances of weighing equities against other values and considering the merits when equities conflict. This volume is dedicated to the important enterprise of raising equity to its proper place as equal to efficiency among criteria to evaluate water-related actions and policies.

Issues of equity are best explored in the context of actual cases of social interaction related to water. Much of what is interesting and important about equity flows from the particular context and longstanding relationships among parties at issue. In virtually every case of water and equity, history is important. The perceptions and behavior of people cannot be understood outside of appreciation for, and recognition of, precedent and place. The chapters in this volume consider water and equity in very different situations with an eye to tracing the ways in which equity arguments are constructed, and the situations in which they are effectively expressed and affect the course of water-related decision making.

We adopt the normative case-study method as an approach to investigate the meaning of values, as advanced by David Thatcher (2006). Just as case studies can contribute to identifying causal relationships and the worldviews of people being studied, normative case studies can clarify important public values. Normative case studies contribute to identifying and understanding the ideals society should pursue and obligations that should be accepted. By choosing cases from different contexts as we have in this book, we can help clarify equity as intrinsic as well as an instrumental value. The case studies are intended to facilitate rethinking the ideal of equity by bringing into view through case studies different circumstances in which the concept was applicable. We draw from a wide range of geographical and historical circumstances that inspire equity judgments, and through juxtaposing cases, reflect on analogies. The case studies allow us not only to refine the meaning of equity but also to trace the moral consequences of ignoring or undervaluing the moral claim in managing water resources, addressing disputes over water, and in decisions over how—and by whom—water will be used.

The final chapter of this volume moves from the particulars of the case studies to some overarching concerns. It looks into the future to consider the impact of global climate change on water resources. The chapter moves beyond the impacts of the melting of polar ice and of sea-level rise to consider widespread changes in the pattern and form of precipitation, and more severe climatic events such as floods, droughts, and high-impact storms. Even the energy policies intended to respond to climate change by finding substitutes for fossil fuels have consequences to water resources, as every recipe for creating energy includes in its instruction, "add water." The example of bio fuels is a case in point as the diversion of water resources to grow crops from energy means depriving present water uses. The burdens of these climate-induced changes are not distributed fairly, and the need to scrutinize equity in future water resources decisions is bound to intensify.

The volume concludes by examining some alternative ways to better incorporate equity into future water resources decisions. An argument is made to move beyond utilitarianism, and the final chapter proposes serious consideration of the advantages and disadvantages of covenants, categorical imperatives, and stewardship. Among the systemic reforms required are inclusiveness and ethical eclecticism, a commitment to democratic processes, making ethical assumptions clear and transparent, and collaboration among parties who disagree. Learning from past error is essential if equity is to be better served, and the final chapter closes with some suggestions for future research.

References

Anderson, Terry L., and Donald R Leal. 2001. *Free Market Environmentalist.* New York: Palgrave.

Arrojo-Agudo, Pedro. 2006. "Water Management in Alpine Regions." Paper presented at the Fifth International Biennial Forum on Water Policy entitled "Managing Upland Watersheds in an Era of Global Climate Change," Banff, Alberta, September 6–11. http://rosenberg.ucanr.org.

Basso, Keith H. 1996. *Wisdom Sits in Places: Landscape and Language among the Western Apache.* Albuquerque: University of New Mexico Press.

Bhatia, R., and M. Bhatia. 2006. "Water and Poverty Alleviation: The Role of Investments and Policy Interventions." In Peter Rogers, M. Ramon Llamas, and Luis Martinez-Cortina, *Water Crisis: Myth or Reality?* London: Taylor and Francis Group, pp. 197–220.

Blatter, Joachim, and Helen Ingram, eds. 2001. *Reflections on Water: New Approaches to Transboundary Conflicts and Cooperation.* Cambridge, MA: MIT Press.

Boulton, L., and R. Sullivan. 2000. "Thirsty Markets Turn Water into a Valuable Export: Drinking Water Is Ever More Scarce in the Parched Lands around the Mediterranean." *Financial Times (London),* November 7, 2000, 15.

Brown, F. Lee, and Helen Ingram. 1987. *Water and Poverty in the Southwest.* Tucson: University of Arizona Press.

Dryzek, John S. 1997. *Democracy in Capitalist Times: Ideals, Limits, and Struggles.* New York: Oxford University Press.

Erie, Steven P. 2006. *Beyond Chinatown: The Metropolitan Water District, Growth, and the Environment in Southern California.* Stanford: Stanford University Press.

Feldman, David L. 1991. *Water Resources Management.* Baltimore: Johns Hopkins University Press.

Gleick, Peter, Gary Wolff, Elizabeth L. Chalecki, and Rachel Reyes. 2002. *The New Economy of Water: The Risks and Benefits of Globalization and Privatization of Fresh Water.* Oakland: The Pacific Institute.

Glennon, Robert. 2005. "Water Scarcity, Marketing, and Privatization," *Texas Law Review* 83, no. 7 (June): 1873–1902.

Ingram, Helen. 1990. *Water Politics: Continuity and Change.* Albuquerque, New Mexico: University of New Mexico Press.

Ingram, Helen, Lawrence Scaff, and Leslie Silko. 1986. "Replacing Confusion with Equity: Alternatives for Water Policy in the Colorado River Basin." In G. D. Weatherford and F. L. Brown, *New Courses for the Colorado River: Major Issues of the Next Century.* Albuquerque, NM: University of New Mexico Press, pp. 177–199. [Reprinted in J. Finkhouse and M. Crawford, eds., *A River Too Far: The Past and Future of the Arid West* (Reno: University of Nevada Press, 1991), pp. 83–102.]

Ingram, Helen and Raul Lejano. Forthcoming. "Collaborative Networks and New Ways of Knowing" in Collaborative environmental policy and adaptive management: Lessons from California's Bay-Delta Program." *Environmental Science and Policy,* special issue.

Maass, Arthur, and Raymond L. Anderson. 1978. *And the Deserts Shall Rejoice: Conflict, Growth, and Justice in Arid Environments.* Cambridge, MA: MIT Press.

Meyer, Michael. 1984. *Water Law in the Hispanic Southwest: A Social and Legal History, 1550–1850.* Tucson, AZ: University of Arizona Press.

Morgenthau, Hans J. 1945. "The Evil of Politics and the Ethics of Evil." *Ethics* 56, no. 1:1–18.

Nichols, John. 1974. *The Milagro Beanfield War.* New York: Holt, Rinehart and Winston.

Oggins, Cy, and Helen Ingram. 1989. "Measuring the Community Value of Water: The Water and Public Welfare Project." Sponsored by the Udall Center for Studies on Public Policy, the University of Arizona, and the Natural Resources Center, the University of New Mexico School of Law. Tucson, AZ: University of Arizona.

Ostrom, Elinor. 1990. *Governing the Commons: The Evolution of Institutions for Collective Action.* Cambridge, UK: Cambridge University Press.

Priscoli, Jerome Delli, James C. I. Dooge, and Ramon Llamas. 2004. *Water and Ethics: Overview.* Paris: UNESCO Series, Essay 1.

Rosenberg Forum. 2006. Fifth Biennial Meeting, "Managing Upland Watersheds in an Era of Global Climate Change," Banff, Alberta, September 6–11. http://rosenberg.ucanr.org.

Rudge, D. 2001. "Delegation going to Turkey to discuss importing water." *Jerusalem Post,* January 17, 2001, 5.

Sandford, Robert W. 2003. *Water and Our Way of Life.* Fernie, British Columbia: Rockies Network.

Schouten, Marco, and Klaas Schwartz. 2006. "Water as a Political Good: Implications for Investments." In Joyeeta Gupta, ed., *International Environmental Agreements: Politics, Law, and Economics.* Netherlands: Springer Dordrecht.

Stein, Robyn. 2000. "South Africa's New Democratic Water Legislation: National Government's Role as a Public Trustee in Dam Building and Management Activities." *Journal of Energy and Natural Resources Law* 18, no. 3:284–295.

Thatcher, David. 2006. "The Normative Case Study." *American Journal of Sociology* 111, no. 6:1631–1676.

Thompson, Barton H. Jr. 2000. "Tragically Difficult: The Obstacles to Governing the Commons," *Environmental Law* 30: 241–278.

UN (United Nations). 1992. *The Dublin Statement on Water and Sustainable Development.* International Conference on Water and the Environment. http://www/wmo/ch/web/homs/documents/English.icwedece.html.

UN (United Nations). 1999. *The State of World Population.* United Nations Population Fund.

UN (United Nations). 2004. *Interim Report of the Task Force 7.* United Nations Millennium Project.

UN/WWAP (United Nations / World Water Assessment Programme). 2003. *Water for People, Water for Life.* UN World Water Development Report, UNESCO, and Berghahn Books.

Vaux, Henry Jr. 1988. "Growth and Water in the South Coast Basin of California." In Mohamed El-Ashry and Diana Gibbons, *Water and Arid Lands of the Western United States.* Cambridge, UK: Cambridge University Press, pp. 233–275.

Young, Robert A. 2005. *Determining the Economic Value of Water: Concepts and Methods.* Washington, DC: Resources for the Future.

I

Water, Place, and Equity

Water has constitutive as well as transformative powers. Landscapes that we treasure, such as Niagara Falls and the Grand Canyon, display the power of water at work both to infuse its watery presence into, and to utterly transform both spectacle and observer. Communities that we care about are built around water in the most mundane sense of needing water to sustain them. Communities are also built around water in the loftier sense of forging an identity. The anthropologist Keith Basso has written about the ways in which place can become shorthand for social ties (Basso 1996). For the Native Americans he studied, the water in situ was an integral part of place. Place names evoked moral lessons and community wisdom. For the authors of the chapters in this part of the book, place is an active participant in their effort to reveal equity and their efforts to nurture, achieve, and sustain it.

Aridity has powerful social consequences. The struggles to cope with the conditions of water shortage can serve to bind a community together. The first chapter in this part, by Clay Arnold, examines how a place, the San Luis Valley in Colorado, was able to develop a "moral economy" that made it stronger and better able to make wise decisions. Water flows downhill in places called watersheds. Chapter 3, by Sheldon Kamieniecki and Amy Below, considers a new aspect of the enduring cleavages that divide communities upstream from communities downstream. The authors discuss how to deal with the inequities that result from contemporary "user pays" efforts to clean up coastal streams in California. Is it fair for the people basking in the sun on the beaches of the Southland, and the merchants who sell to them, to bear none of the costs of cleanup that must take place in less wealthy rural areas and small towns at the headwaters?

International borders make for strange places in which to manage shared waters. Only Moses, not modern nation states, can effectively cleave them apart. In chapter 4, Margaret Wilder considers the implementation of water resources reforms strongly supported by most water resources professionals and international financial institutions in the border state of Sonora, Mexico. In chapter 5, Stephen Mumme recounts the long history of inequity in governance of shared waters between the United States and Mexico. In both chapters the familiar story plays itself out of power asymmetries where the rich get richer and the poor get poorer. Better-off farmers are prospering under new participatory and market arrangements while their poorer neighbors sink lower into poverty. Transboundary water management takes halting steps toward equity, but new incidences of injustice discussed in chapter 5 are likely to further embitter future relations between Mexico and the United States.

Water in plentiful supply does not necessarily lead to happy places. The most abundant freshwater supplies on earth have not delivered from equity conflict the Pacific Northwest , nor relationships between the United States and Canada over shared water. Focusing on the Columbia and the Fraser rivers, Paul Hirt, in chapter 6, finds that the richer border neighbor developing first and most is not necessarily the one better off today. He argues that Canadians learned from the mistakes made by the United States in hydroelectric power development on the Columbia River. The stream of broken promises to indigenous peoples whose fishing rights were ignored on the Columbia River are as sad as the near loss of wild salmon.

References

Basso, Keith H. 1996. *Wisdom Sits in Places: Landscape and Language among Western Apache*. Albuquerque: University of New Mexico Press.

2

The San Luis Valley and the Moral Economy of Water

Thomas Clay Arnold

Distributive criteria and arrangements are intrinsic not to the good-in-itself but to the social good. If we understand what it is, what it means to those for whom it is a good, we understand how, by whom, and for what reasons it ought to be distributed. All distributions are just or unjust relative to the social meanings of the goods at stake.
—Walzer 1983, 9

Water in the San Luis Valley is a story unto itself.
—Halaas 1999

On January 30, 2002, The Nature Conservancy, an organization dedicated to preserving unique and endangered ecosystems, announced one of its more ambitious North American projects—the purchase of the 97,000-acre Baca Ranch, located in Colorado's San Luis Valley, for $31 million. Although a pristine landscape worthy of protection in its own right, The Nature Conservancy sought the Baca Ranch in order to protect the San Luis Valley's famous but fragile Great Sand Dunes Monument, located just outside the ranch's border. Widely hailed at the time for the role the acquisition would play in establishing a much larger Great Sand Dunes National Park, San Luis Valley residents, who first proposed the transaction, rejoiced for what were for them more important reasons. As Governor Bill Owens put it, the purchase "represents much more than land preservation. This finally ends speculation over the transfer of San Luis Valley water to the Front Range. The acquisition represents the permanent preservation of the history, culture, and way of life of the entire San Luis Valley."[1]

This chapter reconstructs and analyzes the dynamics that explain how, in saving sand dunes, Valley residents preserved their identity as

a valley community and how, in staving off large-scale water transfers, they maintained control over their collective affairs. These dynamics confirm what I call the moral economy of water, the fundamental,
at times taken-for-granted, normative principles that ultimately inform
Westerners' determinations of the legitimacy or illegitimacy of existing
or proposed water-related practices, developments, or policies (Arnold
2001). Two principles capture the moral economy of water in the San
Luis Valley. One is the principle of complex equity, a standard for evaluation based on water's multiple spheres of meaning and value. From the
perspective of Valley residents, a perspective reaffirmed through decades
of struggle and conflict, just water policies reflect the nature of the good.
Given water's complex nature, no one sphere of meaning and value justly
trumps all the others. Due process constitutes the second principle. Official (re)allocations of otherwise limited supplies of water are just only
if they are also fair. Water policies are fair when they involve all affected
interests, carefully consider all posed alternatives, and rest on known and
good reasons.

Beyond explaining the San Luis Valley, the moral economy of water
clarifies the social, cultural, and especially political factors that confront the growing calls for significantly more market-driven transfers
and urban-oriented allocations of western water. Successfully adjusting to these new demands will require, as the San Luis Valley illustrates,
policies that honor the substantive and procedural principles of water's
moral economy, a foundation for prescription unlike those commonly
proclaimed or defended in the literature on western water politics and
policies. I develop these substantive and procedural principles in the final
part of the analysis, noting (1) how they explain political conflict and
reform; (2) how they enlarge the concept of sustainability; and (3) how
they reintegrate normative concerns with public policy, overcoming the
often noted "moral austerity" (Gillroy and Bowersox 2002) of western
water resources management.

Attending to the moral economy of water pays additional benefits; it
reveals, if only in a preliminary way, the outlines of a new framework
for analysis, new at least in relation to western water politics. Scholars
increasingly debate the assumptions and principles for explaining water
politics (Blatter and Ingram 2001; Arnold 1996). This chapter contributes to that debate, especially in terms of the analytic principles from
which it proceeds.

Two premises, one theoretical, the other methodological, stand out. The theoretical premise is that of interpretive social science in general: "human beings act toward things on the basis of the meanings that the things have for them" (Blumer 1969, 2). Methodologically, then, explaining human affairs (in this case, western water politics) requires identifying and grasping the meanings that frame the field of reference as well as shape the actions taken. Explanations begin with careful reconstructions of the historical, social, and cultural contexts within which actors operate. Developed below, reconstruction demonstrates that the political history and moral economy of water in the San Luis Valley rest on two key findings. First, water means many different things to residents of the Valley (or, for that matter, to Westerners in general); meanings are as varied and complex as the uses to which water is put. These meanings and uses include but are not limited to the agricultural, recreational, aesthetic, commercial, industrial, and ecological. Efforts to reconcile these competing uses and meanings (both within and outside of the Valley) frequently dominates Valley politics, especially given fluctuating but consistently tight or short supplies of water. The moral economy of water speaks to the manner in which these reconciliations are (or should be) made and the criteria on which they rest. Second, among water's array of meanings, those of community and municipality often prove decisive.[2] They explain, for example, the Valley's recent and surprising string of political victories over ongoing proposals to market Valley water to far more populous and wealthy urban entities. They explain, in other words, the Great Sand Dunes National Park.

The San Luis Valley

The Great Sand Dunes National Park lies within North America's largest and highest alpine desert. Extending from south-central Colorado into portions of north-central New Mexico, the San Luis Valley is nearly the size of New Hampshire (8,193 square miles). Elevations range from 7,500 feet, along the valley's generally flat and treeless floor, to well over 12,000 feet in the surrounding San Juan and Sangre de Cristo mountains. Although a very arid region, receiving on average less than 10 inches of precipitation a year, water resources are surprisingly abundant. The Valley's water resources include approximately 971 miles of streams (including the headwaters of the Rio Grande and Conejos rivers), Colorado's most extensive system of wetlands, numerous artesian wells, several hot springs, and two overlapping aquifers containing an estimated two billion acre-feet of groundwater.[3]

These distinctive physical features have long affected the course and character of activities within the Valley, including its politics. Valley history is the story of residents adjusting to their physical environment, even as they struggle to overcome key aspects of it, aridity in particular. Where early settlers struggled to establish community and society in the desert, today residents face an arguably greater challenge—preserving all that has been achieved. In this contemporary struggle, as much as in the earlier one, residents' conclusions about what is politically necessary, possible, or appropriate turn on how residents view and value water. Significantly, the value residents place on water is not simply or even most importantly an economic one. As Governor Owens's statement indicates, the value of water is much more complex, a matter of Valley history, culture, and way of life.

Water Development
Between 1851 and 1852, Spanish-speaking settlers from New Mexico established the small community of San Luis. Located in the southeastern quadrant of the San Luis Valley, it would become Colorado's first permanent, non-native American settlement.[4] Earlier attempts to establish permanent settlements throughout the Valley failed in the face of numerous hardships, not the least of which was agriculture in a land of very little rain and a relatively short growing season (approximately 90 to 100 days). The people of San Luis turned to irrigation, drawing on well-established practices from New Mexico. In 1852 they filed claims on, and dug a ditch to, the waters of Culebra Creek. Underscoring their collective efforts, residents of the fledgling community named their ditch the San Luis People's Ditch. It remains the oldest continuously used ditch in Colorado. Many other settlements followed San Luis's lead. By the 1860s more than 40 irrigation-based communities (*acequias*) had emerged (Simmons 1999, 86–87, 107–108; Ogburn 1996, 7–8), an early and vivid confirmation of the Hispanic refrain, which Valley residents repeat to this day, *Sin agua no hay vida* ("Without water, there is no life").[5]

Acequias in the San Luis Valley followed a pattern of development and organization typical of the Hispanic Southwest, a pattern centered on meeting communal needs through common enterprise. Residents played a key role, beginning with irrigation. Often community members hand dug the vital waterworks, outlining in the process the social and political, as well as physical, contours of their developing association.[6] Within the sphere of inhabitation afforded by the main ditch (*acequia madre*) and its laterals, settlers built plazas, erected churches, established missions or schools, and platted public and private lands. Through the collective provision of water, settlers won the rural yet social mode of existence they could not win or enjoy in any other way.

Generations removed from the rigors of establishing ditch communities, Valley acequias remain strikingly communal associations, with members bound to one another in relationships still formed around and sustained by the mutual administration of water. These simultaneously social and political arrangements, among the oldest mechanisms for local governance in the United States (Hicks and Pena 2003, 164), account for a substantial portion of the culture and way of life Governor Owens declared preserved by The Nature Conservancy's purchase of the Baca Ranch.

The communal association is headed by the *majordomo* (ditch boss), an elected member of the acequia who supervises, sometimes in conjunction with an elected community ditch commission, the repair and annual cleaning of the ditches (drawing on the labor members are obligated to contribute). The majordomo also enforces agreed upon rules, fines violators, resolves disputes, and, most importantly, designs an irrigation schedule that assures each user a turn, even in times of scarcity. Majordomos craft their schedules and allocations by taking into account, among other things, the total amount of water available, the crops in question, the amount of land owned and to be irrigated, need, and the common good (Rivera 1998, 18–19, 38, 56–58). Feasible and legitimate schedules depend on majordomos and members working together, exchanging relevant information, and sharing perspectives, reproducing in the process customary ways and traditional ties. Significantly, members of the Valley's acequias understand themselves predominantly in terms of their connections to each other, to the community ditch, and to their role in meeting community needs.[7] For these residents of the San Luis Valley, water is not simply a necessary ingredient in the cultivation of crops. Water also signifies a sense of attachment and mutual obligation by which they recognize themselves as members of a community. Members of acequias value their water accordingly, as the lifeblood of the community.

Valley history and culture also reflect the influx of non-Spanish speaking colonizers. Anglo, Dutch, Swedish, German, and Mormon settlers (hereafter collectively referred to as Anglo-American) flocked to the Valley in the years following the 1862 Homestead Act. Compared to their Hispanic predecessors, these settlers brought with them a much more commercial spirit, prominently displayed in their development of ranching, mining, railroads, and retail industries. Most, however, turned to irrigation-based agriculture, reproducing in their own way a fairly communal view of water.

In the last decades of the nineteenth century, communities, agriculture, and irrigation increased sharply in the San Luis Valley, lifting the Valley out of its pioneer and frontier stage. Mormons established the communities of Romeo, Eastdale, Sanford, Ephraim, Richfield, and Manassa, while Dutch settlers founded La Jara, Waverly, and Bowen, all in the south-central reaches of the Valley. Swedes, Germans, and Anglos established numerous communities in the northern, central, and western sections, including the prominent communities of Alamosa, Del Norte, Creede, Saguache, Mosca, Hooper, and Monte Vista (Simmons 1999, 217–225).

Sustainable agriculture and communities coincided with a "ditch boom." Between 1880 and 1890 alone, canal companies constructed six ditches capable of irrigating between 300,000 and 500,000 acres of farmland. The Rio Grande Canal was the largest. Containing approximately 210 miles of ditches and laterals, it dwarfed many of the earlier Hispanic acequias. Three to five thousand men using a thousand teams of horses constructed the main canal between 1881 and 1884. Today it diverts approximately 30 percent of the Rio Grande and serves 517 different headgates across three different counties in the Valley (Ogburn 1996, 9, 10).

Although the Rio Grande Canal is the Valley's largest, it was but one part of an even larger irrigation and community-building project, T. C. Henry's Del Norte Land and Canal Company. In the early 1880s, Henry purchased two large tracts of land (7,000 acres total) north and south of what was to become Monte Vista. The tracts were to be divided into smaller farms and sold at three to five dollars an acre. Success rested on the availability of water. The Rio Grande Canal was one of several ditches built to service what Henry called the North and South Farms. His network of canals included the San Luis Valley Canal, Empire Canal, and Citizen's Ditch. Henry's finances failed him in 1885, but Monte Vista and the surrounding area successfully established itself as one of Colorado's most productive agricultural regions.[8] Similar but smaller developments occurred in and around Mosca, Hooper, and Alamosa. Small farming communities, often consisting of a church, school, mill, and train depot, proliferated. Even before the turn of the century, however, many residents lived in far more developed civic associations. Monte Vista, for example, boasted a "three-story hotel . . . a roller and an elevator; warehouses; a machine shop; creameries; banks; newspapers; a public library; six churches," and, in 1915, a sugar-beet factory (Simmons 1999, 229).[9]

Anglo-American irrigation in the Valley featured its own form of cooperative administration, the mutual irrigation or ditch company. Fervently supported by Henry (see Sherow 1990, 12, 18), mutual companies were founded as (and remain today) voluntary associations whose "purpose is the distribution of water, not profits, to its shareholders" (Corbridge and Rice 1999, 283). Members deeded their water rights to the company for shares of stock, issued in proportion to the acres of land irrigated. Most ditch companies in the Valley today are incorporated. They operate according to written bylaws, elect boards of directors, hold regular meetings, "levy assessments either in money or labor for the construction and maintenance of the project," and employ watermasters (also known as ditch riders) "to distribute the water equitably and fairly among the shareholders" (Dunbar 1983, 29).[10]

Although perhaps not as intensely communal as the Valley's venerable acequias, mutual ditch companies foster similar views and practices.[11] Like acequias, mutual companies rely on members for maintaining their joint endeavors and managing their collective affairs. In mutual companies, as well as in acequias, members form and then act in light of what Hicks and Pena (2003, 193) call "enforceable networks of reliance and of mutual expectation." Like acequias, mutual companies limit "the exercise of individual water rights in the interest of common ends" (ibid). Equally important, at least from the perspective of establishing the basis for an emerging moral economy of water, members and residents alike understand water as an intrinsically communal social good. In the arid San Luis Valley, water signifies what its provision both requires and sustains—participation with others in a common life.[12] When understood and valued in these terms, water has the potential of becoming the central frame through which residents both interpret and fashion responses to unfolding events.[13] Persistent and Valley-wide struggles related to the Rio Grande Compact realized that potential. Subsequent events, in particular those now known as AWDI and Stockman's Water, confirmed it.

Rio Grande Compact

By the 1890s, irrigation in the San Luis Valley had grown to the point that the Rio Grande was delivering noticeably less water downstream. The older irrigation operations and communities of the Mesilla and El Paso–Ciudad Juarez valleys suffered accordingly. Matters deteriorated in 1893 when the entire region battled both financial collapse and an especially severe drought. Acrimony over the river's rapidly dwindling flows escalated. Residents of New Mexico and Texas petitioned for a large reclamation reservoir, contending bitterly over where it would be located. The Republic of Mexico threatened legal action against the U.S. government. The government responded in 1896, placing an "embargo" on further irrigation-related water development in the upper Rio Grande, an embargo that, in one form or another, lasted until the early 1920s. Negotiations with Mexico culminated in a 1906 treaty, guaranteeing Mexico 60,000 acre-feet of Rio Grande water a year (Paddock 2001).

Residents of the San Luis Valley viewed the embargo negatively. Younger and less developed than the downstream valleys, the San Luis Valley would bear the brunt of the embargo's restrictions. Residents chafed at not being able to construct reservoirs and other diversion projects they believed vital to the security of their hard-won but now, once again, tenuous social, political, and economic achievements. Drought and economic depression alone had already caused a number of banks, farms, and even small communities to collapse (Simmons 1999, 231); the embargo, residents believed, would only make matters worse. Anxious to remove the embargo, residents pushed hard for an interstate compact to settle once and for all allocations between New Mexico, Colorado, and Texas. The result was the Rio Grande Compact of 1938. According to the Compact, the State of Colorado (in effect, the San Luis Valley) must deliver specific, but fluctuating, amounts of water in the Rio Grande where it enters New Mexico. The required amounts vary based on the total amount of water available, and as measured by several Compact-required gauging stations. Wetter years increase Colorado's obligations, while drier years lower them.

Meeting these requirements proved difficult, especially through the 1950s and 1960s. By 1965, Colorado was 1 million acre-feet short on its obligations, in part due to several wet years, as well as the tremendous increase of groundwater wells in the San Luis Valley.[14] New Mexico and Texas sued in 1966. Colorado subsequently agreed to sharply reduce diversions of water in the San Luis Valley, reviving and permanently establishing residents' concerns about maintaining their identity as a valley of viable social and political communities in the face of outsiders interested in Valley water. In addition to capping artesian wells and restricting the drilling of new wells, the state engineer of Colorado ordered reductions in the amount of water used in surface irrigation. Between 1968 and 1984, these reductions amounted to 890,000 acre-feet of water (Ogburn 1996, 21). The traditional practice of irrigating some fields in late winter, to bring up the water table before spring planting, was eliminated.

Residents sought relief from these burdens, as well as from the specter of a federal water master, by increasing their capacity to act as a unit. In 1967, Valley residents petitioned the state for the authority to form a Valley-wide tax and water conservation district, explicitly for the purpose of representing the San Luis Valley in matters pertaining to its Rio Grande Compact obligations. The state granted the authority but stipulated that the formation of a Rio Grande Water Conservation District had to be put to a popular vote. Residents responded 13–1 in favor of the district, a particularly striking case of municipality for the sake of community.[15]

Conservation district officials and Valley residents pinned their hopes on resurrecting a project first proposed in the 1920s—augmenting Rio Grande flows with water from the Valley's Closed Basin. Often called the "sump," the Closed Basin is a nearly 3,000-square-mile watershed north and east of the Rio Grande, a region that includes the Baca Ranch and the Great Sand Dunes. Surface and ground waters flowing into the Closed Basin cannot, given its geology, naturally flow out, making it the repository of a significant amount of water, particularly groundwater. In 1972 the Rio Grande Water Conservation District filed a petition for rights to Closed Basin waters otherwise lost to evapotranspiration by various forms of vegetation. Congress subsequently authorized the Closed Basin Project as a federal reclamation project. Constructed between 1981 and 1993, the project consists of 170 groundwater wells pumping water to the Rio Grande via a 42-mile conveyance channel.

Early and continuous struggles over the disposition and control of the Valley's waters, famously symbolized by the embargo and Rio Grande Compact, have affected the Valley in several ways. Residents realize that, despite the recourse to ever-more sophisticated forms of water development, they have not escaped the prospect of a ruinous shortage of water. Each generation relives in its own way the hopes and fears of the Valley's original settlers but with this added factor: in addition to

the challenges of naturally occurring aridity, residents must now contend
with the equally daunting prospect of shortages produced by external
calls on Valley water. Consequently, Valley residents, even though they
live and work in many different and widely scattered communities, think
of themselves, at least when it comes to water, as stakeholders in and advocates for a larger, encompassing entity—the San Luis Valley.[16] Given
the long-term and interstate nature of the challenges related to the Rio
Grande Compact, water is the most important way in which residents
think about themselves, their communities, and, as one observer put it,
"their relationships to the rest of the state and the West" (Hughes 1998,
H-12). Due to the struggles surrounding the Rio Grande Compact experience, Valley residents have retained an especially keen sense of what is
ultimately at stake should other entities outside the Valley seek and obtain large, annual withdrawals of water. Significantly, the stakes involved
go beyond the relatively meager monetary returns of their predominantly
agricultural economy.[17] The stakes also include the identities, relationships, and civic associations long tied to the provision and use of the
Valley's waters. These sentiments galvanized the Valley when, in 1986,
ninety years after the hated Rio Grande embargo, the American Water
Development, Inc. (AWDI) filed claims for 200,000 acre-feet of Valley
groundwater.

American Water Development Inc.   American Water Development Inc.
(AWDI) was the corporate instrument of Maurice Strong, the Canadian
oilman, industrialist, undersecretary for the United Nations, and, from
1978 to 1995, owner of the Baca Ranch, located immediately to the
north of the Great Sand Dunes. Strong founded AWDI to develop the
enormous economic potential of the waters beneath his ranch. Through
AWDI, Strong proposed drilling more than a hundred wells and selling
the water to thirsty cities along the Front Range (Denver, Pueblo, and
Colorado Springs) as well as in the states of New Mexico, Texas, and
California. AWDI's board of directors included several influential individuals drawn to the project: Richard Lamm, former three-term governor
of Colorado; William Ruckelshaus, past director of the Environmental
Protection Agency; and Canadian entrepreneur, Saul Belzberg, CEO of
First City Financial Corp. of Vancouver.

Valley residents responded with alarm and anger, denouncing the proposal as "a tremendous threat to their well-being and continued existence" (Ogburn 1996, 30). Drawing on the sentiments and institutions

extruded from generations of struggle over water, Valley residents quickly committed to the fight, waging it on two fronts. On the legal front, 154 different entities within the Valley (led by the Rio Grande Water Conservation District, product of an earlier water war) combined forces to contest AWDI's petition in court. To meet their legal expenses the Rio Grande Water Conservation District proposed borrowing money from local lenders, using extended property tax obligations as collateral. Of the nearly nine thousand votes cast on the proposition, 98 percent favored the tax increase (Gascoyne 1991, 8).

On the political front, residents formed a potent grassroots organization, Citizens for San Luis Valley Water (CSLVW), to publicize their plight, promote their cause, and mobilize opposition to AWDI by passing out bumper stickers, holding rallies,and organizing fundraising events. Members traveled the state with a large mural depicting the people of the Valley rising up to protect their resources. The group also lobbied the legislature and started *Valley Voice,* CSLVW's widely read periodical dedicated to informing readers about AWDI and related water affairs (Canaly 1997, 13). Supported by Hispanics and Anglos, farmers, ranchers, and environmentalists, CSLVW's efforts exceeded expectations. Journalists covering the AWDI trial observed that even Valley police cars carried anti-AWDI bumper stickers (Noreen 1994). The BBC featured the Valley in a documentary entitled "Water Wars," spreading the Valley's story well beyond the boundaries of Colorado. AWDI's petition was denied in 1991, a denial eventually upheld by the Colorado Supreme Court.[18]

CSLVW's success rested on several factors. One was the Valley's history of water-related community, municipality, and coalition, easing the transition to even more focused forms of collective action.[19] Similarly, CSLVW did not have to build vital networks of water-oriented association and communication completely from scratch, saving them time and energy for other equally important anti-AWDI activities. A second, closely related factor, was the magnitude of the threat symbolized by AWDI. Nearly every sector of the Valley stood to suffer, critics alleged, if AWDI won its petition. Removing an additional 200,000 acre-feet of groundwater a year from the Closed Basin would, CSLVW and others concluded, jeopardize irrigation-based agriculture and grazing, as well as endanger the flora, fauna, and wildlife of the Closed Basin's wetlands, destabilize the Great Sand Dunes (held in place by the groundwater moisture at their base), and imperil the Closed Basin Project, the critical groundwater component in the Valley's Rio Grande Compact plans.

Shared threats, especially if thought large and ominous, often override familiar but lesser points of internal difference and friction. So, too, with AWDI and the San Luis Valley; the all-encompassing quality of the threat unified residents even as it angered and alarmed them.[20]

Characteristic of those informed by a moral economy, opponents focused their efforts on demonstrating the injustice of AWDI's proposal. Residents' perceptions of the injustice at play developed quickly, especially given the two cases of water exportation to which AWDI was most often compared: Owens Valley, California and Crowley County, Colorado (Saunders 1991). Their circumstances mirrored those of the San Luis Valley—rural, largely agricultural, dependent on irrigation, and the target of cities seeking additional sources of water. Unable to counter the power of the cities, Owens Valley and Crowley County lost their water. Their once viable regions slowly transformed into social, economic, and environmental wastelands.[21]

Presented with these comparisons, residents drew the obvious conclusions: ruin and injustice lay at the heart of AWDI's proposal.[22] It is significant that residents did not restrict their claims of injustice to the likely long-term effects of exporting an additional 200,000 acre-feet of Valley water. An equally galling injustice resided in AWDI's conception of water as nothing more than an article of commerce and speculation. Residents rejected this economic view of water on two grounds. First, water is not merely a commodity for exchange. It is also a complex social good, an integral part of the Valley's culture and way of life. Furthermore, water is the foundation for values (e.g., identity and community) unrelated to price, and properly considered not for sale.[23] Second, commodifying water ultimately shifts the responsibility for Valley affairs (so often centered on water) from those most directly affected and traditionally involved (Valley residents) to those with wealth, even if, as in the case of Strong and Belzberg, they are citizens of a foreign country.[24] In short, Valley residents found AWDI illegitimate. They found it a violation of their historical understandings of the role, place, and purpose of water in collective affairs, sentiments and understandings they would invoke even more fervently against "the son of AWDI," Stockman's Water.

Stockman's Water   Saddled by defeat and rising legal costs, Strong sold the Baca Ranch in 1995 to Gary Boyce, a recently returned native of the Valley. Boyce purchased the ranch for $16.5 million, with the bulk of the money coming from Yale University and Farallon Capital Management, a billion-dollar investment firm in San Francisco.[25] Boyce and his backers wanted the Baca Ranch to succeed where AWDI had failed—selling groundwater to thirsty western cities. Struck by the prospect of selling water at the prevailing rate of $4,000 to $7,000 an acre-foot, Boyce announced that he and his Stockman's Water Co. were "in the San Luis Valley to do business."[26]

Like AWDI, Stockman's plans featured drilling numerous wells and selling the recovered water, up to 100,000 acre-feet, to metropolitan Denver and Colorado Springs, perhaps even to cities in New Mexico and California. Success hinged, Boyce concluded, on avoiding the kind of outrage AWDI had incurred, on avoiding the perception that developing Baca Ranch groundwater would harm the Valley. Boyce adjusted his proposal accordingly, promising to:

• Use, if necessary, his ranch's senior surface-water rights (25,000–50,000 acre-feet) to compensate for any losses to the Closed Basin.
• Create a $3 million trust fund, to assist anyone needing to drill deeper wells because of Stockman's withdrawals;
• Set aside 50,000 acres for a wildlife reservation.
• Establish a 10,000-acre preserve around his ranch's two fourteen-thousand-foot mountains, Kit Carson Peak and Crestone Mountain.
• Deliver 10,000 acre-feet of water a year for ten years to the Arkansas River, assisting Colorado in meeting its Arkansas River Compact obligations.[27]

Boyce's effort to portray Stockman's as benign, as a "huge conservation project," failed.[28] Opponents sounded the alarm, led by CSLVW. In October of 1996, residents took out a large ad in *The Denver Post*, entitled "Water: Take it . . . or Leave it?" Bearing the names and signatures of several hundred Valley residents and organizations, it reaffirmed their hostility to purely market-based transactions: "We, the undersigned citizens, are committed to preserving the water within Colorado's San Luis Valley for land and life. Private marketers are drawing up plans to take water from the Valley and sell to other parts of Colorado or other regions. *We are opposed to these plans. . . .* We are committed to fight efforts to take water from this beautiful place, the San Luis Valley. Help us protect land and life."[29]

Struck by both the depth of the Valley's opposition and how close Valley representatives had come to passing HB 1214 (which would have effectively prohibited groundwater transfers out of the Valley), Boyce altered his tactics. In 1998, he introduced two constitutional initiatives, amendments 15 and 16, designed to, as he put it, "level" the playing field.[30] Amendment 15 required the installation of flow meters on approximately 3,500 wells in the Valley, presumably to assure both efficiency and conservation in water usage. Amendment 16 required a $40 an acre-foot tax (retroactive to 1987) on waters drawn from beneath state lands within the Valley, to be paid by the Rio Grande Water Conservation District. Amendment 16 directed that a substantial percentage of the revenues collected be earmarked for San Luis Valley schools.

Valley residents viewed Boyce's amendments very differently. Flow meters, they argued, clogged easily, failed often, and cost between $700 and $1,200 each, a prohibitive amount of money for hard-pressed irrigators.[31] The state engineer of Colorado, residents noted, preferred a far more reliable and inexpensive alternative (monitoring power consumption), why not Boyce? Residents found the fact that amendment 15 did not apply to any other region of Colorado equally suspicious. When coupled to the provision exempting Boyce's planned wells from the meter requirement, amendment 15 struck residents as deliberately punitive.[32]

Amendment 16 angered residents even more. Passage meant an immediate tax bill of $8 million to the Rio Grande Water Conservation District, well beyond available resources or, for that matter, the District's tax base (Lofholm 1998, A-01). Moreover, given the intricacies of Colorado's Taxpayers' Bill of Rights (TABOR), only a fraction of those tax revenues would actually reach the Valley's schools.

Boyce's strategies marked a sharp break from the AWDI episode. Residents feared the challenge of a state-wide vote and that a majority of those voting on the amendments would either: (a) fail to see the amendments for what they truly were; or (b) reside in areas most likely to gain from Boyce's export plans. Residents viewed Boyce's amendments as nothing less than an attempt to bankrupt the Valley, rendering it incapable of contesting Boyce in court once he filed for water rights.[33]

Boyce's amendments presented other difficulties. Valley residents would have to carry the fight across the breadth of Colorado, a scale of endeavor unlike earlier, more familiar water contests. Moreover, mobilizing state-wide opposition would require developing a larger repertoire of action, all while fending off a hostile redistricting proposal.[34] Finally, neither the Rio Grande Water Conservation District, a tax-supported entity, nor CSLVW, a not-for-profit organization, could spend money on defeating the two ballot measures.

To raise money for the fight and coordinate action, residents formed Citizens for Colorado's Water. In addition to traditional grassroots activities, Citizens for Colorado's Water turned to state-wide political ads—three 30-second TV spots and three radio ads, one of them in Spanish. One TV ad portrayed Boyce and Farallon as "out-of-state speculators," and amendment 16 as a "water grab initiative" (Abbott 1998, 7A).

Citizens for Colorado's Water succeeded in making the ballot measures an issue of justice versus greedy subterfuge. Numerous state-wide organizations joined in the opposition, among them, the state-wide organization of school executives, Colorado's largest teachers' union, the Colorado Water Conservation Board, the League of Women Voters, the Sierra Club, the Colorado Association of Realtors, Ducks Unlimited, the Colorado Farm Bureau, Trout Unlimited, and the Colorado Council of Churches (ibid). A representative for the Colorado Association of School Executives described Boyce's amendments as an "unconscionable use of Colorado children as pawns in a scheme to further the financial interests of wealthy investors" (Sanko 1998, 36A). Members of the Colorado Council of Churches labeled the whole affair as "dishonest," a matter of "rich and powerful outsiders trying to bankrupt struggling farmers in the San Luis Valley."[35] Even the op-ed page of the Denver-based *Rocky Mountain News* described Boyce's amendments as an "unprincipled scheme," one that could easily "destroy the long-established patterns of water use in the San Luis Valley."[36] Voters responded accordingly, defeating Boyce's amendments by a margin of 3 to 1.[37]

Great Sand Dunes National Park   Victories over AWDI and Stockman's Water demonstrated a sobering truth. As Ralph Curtis, the general manager of the Rio Grande Water Conservation District, put it, "If you leave Baca Ranch in private hands, the owners will certainly try to export the ranch's water" (Lloyd 2000, 2). As in so many earlier cases, residents formed an organization, Citizens for Monument to Park Conversion, to press their cause—public ownership of the Baca Ranch. Senator Wayne Allard and Representative Scott McInnis quickly agreed to introduce legislation authorizing national park status, but only on the condition of federal ownership of the Baca Ranch. Secretary of the Interior Bruce Babbitt embraced the proposal, declaring to the delight of Valley residents, "These water-export schemes have just acquired a new adversary, and the people of the San Luis Valley have a new ally against these schemes" (ibid). President Clinton signed the measure on November 22, 2000. Conversion was assured in January of 2002, when The Nature Conservancy, pledging to donate the property to the federal government, announced its $31 million purchase agreement. Representative McInnis observed the occasion with an announcement of his own. Testifying to the forces behind the conversion, McInnis concluded, "With the Baca, the Great Sand Dunes National Park and Preserve will be second to none in natural beauty and ecological diversity. Equally important, the purchase puts an end to the San Luis Valley's water wars once and for all."[38]

The Moral Economy of Water   Cases like the Rio Grande Compact, AWDI, Stockman's Water, and the Great Sand Dunes illustrate how deeply residents' historical understandings of water have affected Valley politics. Residents' vigorous and generally successful opposition to various efforts to appropriate, transfer, and sell Valley water testifies to a normative consensus embedded in those understandings: a consensus on what distinguishes legitimate from illegitimate policies and practices. The two principles of complex equity and due process lie within that consensus and constitute the moral economy of water. The growing transboundary challenges of water in Colorado and the arid American West underscore the importance of policies and processes that take the moral economy of water fully into account.

The normative consensus residents invoked throughout the events leading to the formation of the Great Sand Dunes National Park stems from water's status as an intrinsically communal social good. In the arid San Luis Valley, water remains the basis for the relationships and commitments, the sense of shared fate and purpose that make associated residents a community. As AWDI and Stockman's Water illustrate, proposed courses of action that do not honor water's status as an intrinsically communal social good or, worse, jeopardize the integrity of the Valley's existing community are illegitimate and rightfully opposed. Justice is in these instances a matter of Valley residents ordering their collective affairs, of maintaining the kind of jurisdiction over water their collective affairs and welfare require.[39]

Significantly, the moral economy of water does not automatically translate into opposition to non-Valley uses of Valley water. Residents' support for the Closed Basin Project (which provides Valley groundwater for downstream and out-of-state Rio Grande water users) indicates a more complex formulation. So do the comments of Ken Salazar, fifth-generation Valley native and U.S. senator from Colorado, made at the height of the battle against Stockman's Water. Although strenuously opposed to Boyce's initiatives and his plans for marketing Valley water to metropolitan Denver (they "would have a devastating effect on the San Luis Valley"), Salazar nevertheless concluded, perhaps with the example of the Closed Basin Project in mind, that "there are cooperative solutions we can fashion to make sure that we provide a viable water supply for the growing communities of the Front Range."[40]

These examples clarify the moral economy of water. They reveal that the grounds for residents' moral and political indignation lie far more in the unnegotiated and unilateral qualities of the AWDI and Stockman's Water proposals than in the prospect of non-Valley uses of Valley water per se. Grounds like these point in turn to the principle of complex equity.

Complex Equity

Complex equity is a form of distributive justice, a regulative ideal for goods distinguished by their multiple meanings and values. According to the principle of complex equity, justice rests on managing complex social goods in light of their multiple meanings and values, and on rendering those meanings and values their due. What are complex social goods and their multiple meanings and values due? Complex social goods merit, first, official recognition of their varied meanings and values within all effective decision-making arenas related to the good in question. Decisions and decision-making processes that fail to regard complex social goods as complex commit a kind of injustice; they conceive the good as something other and less than what it is for many of the individuals and communities most directly affected. More precisely, they conceive the good incompletely and, therefore, unfairly.

Second, beyond the justice of recognizing complexity, complex social goods merit the protection of their multiple meanings. No one particular meaning defines a complex good. Consequently, just arrangements sustain the complexity to which all aspects of a complex good otherwise contribute.[41] Even though it is not always possible to achieve each meaning and value fully, especially if, as with water in the arid West, the good is scarce, meeting one aspect of a complex good need not, and should not, be done at the permanent sacrifice of another. Complex equity opposes reductionism.

Third, complex equity furthers the justice of nondominance. Each meaning within a complex good constitutes it own sphere of value and significance. Conceivably, each sphere entails its own inherent principle of distribution, whether that principle is one of exchange, need, or desert. The regulation of complex goods on the basis of only one inherent principle of distribution leads to a kind of tyranny. It leads to the domination of every other sphere of value and significance by alien, perhaps even hostile criteria. Justice is in these complex cases a matter of preserving each sphere and principle of distribution, by limiting their ability to govern the good in its entirety (Walzer 1983).

AWDI and Stockman's Water violated the principle of complex equity on every count. Their essentially market-driven approach unjustly reduced water to little more than a commodity for self-interested pecuniary gain. Residents responded, as we have seen, emphasizing the communal, municipal, and environmental dimensions of water most at risk.

Due Process

Water in the San Luis Valley is as scarce as it is complex, compounding the basis for moral-economic evaluation. Given water's scarcity and complexity, water policies cannot help but favor some meanings and values over others. From the perspective of what is to be evaluated, decision-making procedures are as important as the decisions themselves. The principle of due process is the claim that decisions must be made fairly.

Fair decisions take place in open, essentially public forums. Open forums grant all affected interests a voice in the decision-making process; they guarantee that all of water's many uses, meanings, and values are taken fully into account. Beyond the benefits of inclusion, participation, and representation, the principle of due process assures participants that the inherently collective act of allocating water remains the responsibility of the interests and communities affected.

Decisions are also fair if grounded in deliberation. The fact that decisions inevitably favor one set of meanings and values over others does not exempt them from the justice of giving an account. Fair decisions rest on known and good reasons. Good reasons demonstrate the superiority of the suggested course of action, a matter of articulating and then carefully weighing the effects of promoting one set of meanings and values rather than another. Successful deliberation rests on meeting all posed objections, and on adhering to processes that allow objections to be raised.

AWDI and Stockman's Water rejected the principle of due process, a critical factor in their defeat. Both proposals assumed, rather than demonstrated, the superiority of water's status as an article of commerce.

Neither proposal embraced as fully and sincerely as they should have the justice of open and public forums, preferring instead the much more private and restrictive sphere of allocations based on wealth. Indignant over their unfairness, residents defeated both proposals.[42]

Conclusion
Western water policy labors under at least three distinct but closely related burdens:

• Growing uncertainty and political conflict over how best to (re)allocate western water.
• Achieving sustainability.
• Reintegrating politics and ethics.

As revealed in the San Luis Valley, the moral economy of water assists on all three fronts. I conclude with a brief description of how and why.

Political Conflict   The moral economy of water clarifies the grounds for some kinds of political conflict, a critical step in designing more effective policies. For residents of the San Luis Valley, and, perhaps, many other westerners, water is not simply a natural resource for ever-more efficient or profitable consumption, the underlying assumption of many market-related policy proposals. Water is much more. It is also a social good, the basis for, among other things, uniquely valued identities, relationships, and civic associations. Policies that ignore or downplay one or more of the many uses, meanings, and values of water increase the chances for political conflict and defeat. Policy success is instead a matter of inclusion and balance. According to Dennis Murphy, former mayor of the San Luis Valley community of Del Norte, "water decisions must be made with [the] full consideration of the long-term and potentially irreversible economic, social, and environmental consequences which those decisions have on our various communities" (McAvoy 2002). In 2003, the Colorado state legislature agreed, passing House Joint Resolution 03–1019, a series of water planning principles, five of which directly reflect the moral economy of water:

• All Colorado water users must share in solving Colorado's water resource problems [the principle of due process].
• In the event that agricultural water is transferred, the transaction must adequately address the need for maintaining the existing tax base, protecting the remaining water rights in the area, and maintaining the proper

stewardship of the land, including revegetation and weed control [the principle of complex equity].

• Adverse economic, environmental, and social impacts of future water projects and water transfers should be minimized; unavoidable adverse impacts must be reasonably mitigated; all communities involved should commit themselves to identifying and implementing reasonable mitigation measures as an integral part of future water projects or transfers [the principle of complex equity].

• Future water supply solutions must benefit both the area of origin and the area of use [the principle of due process].

• The goal of all parties should be to ultimately advance the economic, environmental, cultural, and recreational health of all Colorado communities [the principle of complex equity].[43]

Sustainability   According to numerous accounts, current patterns of western water use cannot continue; they exceed even optimistic estimates of future water supplies and capacities. Policy makers and scholars alike now tout the idea of sustainability but differ over what that idea means or entails. Many favor various notions of sustainable development, which the United Nations World Commission on Environment and Development defines as that level of economic growth which "meets the needs of the present without compromising the ability of future generations to meet their own" (WCED 1987, 43). Others (MacDonnell 1999, 326, n. 21) favor "environmentally compatible development." Under this version, sustainable policies minimize or avoid "environmental losses" (p. 238).

Given its emphasis on what water means for those who use it in the arid West, moral-economic analysis recommends a different approach, factoring in what the two previous definitions leave out—water's status as a complex social good. As established above, no one use or meaning completely defines a complex good. In the San Luis Valley, water is simultaneously the lifeblood of the community, an agricultural commodity, an environmental resource, and more. Complex goods require an equally complex take on sustainability, one centered on sustaining meanings (including communal and municipal meanings) as much as uses. This need not lead to the conclusion that sustainability is nothing more than the stubborn preservation of the status quo. Sustaining complexity is not inconsistent with changes in practice or emphasis. Change, however, must take place within the parameters of the idea of complexity itself. When it comes to complex goods like water, sustaining every other meaning and value is the condition for emphasizing any one of them.[44]

Politics and Ethics One of the more serious complaints about water resources management is that it lacks a principled foundation from which to chart policy. According to David Feldman (1991, 8), water policies instead reflect a "disjointed utilitarianism," an ethically indefensible propensity for satisfying the short-term and largely material interests of those he describes as "vocal enough or well situated enough to make themselves heard." The results, he concludes, are as predictable as they are costly—public policies inconsistent with "justice and rationality" (p. 3). Gillroy and Bowersox (2002, 1) attribute environmental injustice and irrationality to an inherent and persistent "moral austerity," their term for policy makers' inability to meaningfully debate the question, "What is good environmental policy?" In both accounts, progress rests on bridging the wide divide between ethics and politics, on designing a framework that defines "what makes policy 'good' or 'bad,' 'right' or 'wrong'" (Bates et al. 1993, 178).

Moral-economic analysis offers one such framework. Just policies, it asserts, advance the normative principles embedded in our understandings of the goods themselves. Multiple meanings and understandings complicate the analysis, but they do not disprove the point. With respect to water in the arid West, just policies are those that comply with the principles of due process and complex equity.

Notes

1. PR Newswire, January 30, 2002.

2. In this context, *community* means the sense of belonging and related identity brought about by and embedded in shared, ongoing circumstances and experiences. *Municipality* refers to the institutions and mechanisms by which a community conducts its public affairs, often a factor in reproducing senses of community.

3. An acre-foot of water is almost 326,000 gallons, what it would take to cover one acre of land to a depth of one foot. The San Luis Valley's two billion acre-feet of groundwater dwarfs the combined storage capacity (55.5 million acre-feet) of Lake Powell and Lake Mead, the two largest man-made reservoirs in the United States.

4. The San Luis Valley was a part of the Territory of New Mexico until 1861, when it was included in the newly formed Territory of Colorado.

5. A stronger, darker version of this adage comes from Robert Ogburn, the long-term Water Judge for the San Luis Valley and Chief Judge of Colorado's 12th Judicial District. Having presided over 2,500 water cases, he concluded: "Without water, our land is a corpse, whispering soil into the wind, whispering enmities against the human race" (Ogburn 1996, 5). While speaking at a 2005 symposium on the San Luis Valley's groundwater, Colorado Supreme Court Justice Gregory Hobbs offered an even more telling version, at least in reference to the Valley and the moral economy of water. Commenting on the Valley's and Colorado's long struggle with aridity, he concluded: "In scarcity lies the opportunity for community" (Heide 2005).

6. Early frontier settlements in New Spain were most often called *acequias de común,* meaning that the ditch bringing the life-giving waters was "of community" (Rivera 1998, 52, 227). This accounts in part for why the term *acequia* means both the ditch carrying the water (as in *acequia madre*) and the civic association organized around it.

7. Hicks and Pena (2003, 161, n. 187) make the same point but emphasize the longstanding obligation of members to maintain the community ditches. The obligation "to contribute labor," they write, "remains an expression of . . . continuing commitment to a vision of community. The annual spring ditch cleaning is an occasion defined by ritual and festival. The assertion of these commitments is significant as a foundation of community solidarity in a landscape of newly individualized water rights, reaffirming commitment to *acequiadad* (*acequia*hood) and the community of labor that sustains it."

8. In the late 1880s, featured crops included potatoes, wheat, and hay. Today, the San Luis Valley is a major producer of potatoes, barley, carrots, and lettuce.

9. A Denver and Rio Grande Railroad promotional for Monte Vista captured residents' sense of civic and social achievement: "No Eastern town," it concluded, "is more law-abiding or has a higher moral tone. No gambling dens or liquor saloons have ever existed in Monte Vista" (Simmons 1999, 168).

10. According to Pisani (1992, 101), farmers formed mutual companies "to promote efficiency and democracy in the distribution of water." Farmers' shares of stocks "corresponded to the value of their property in relation to that of the whole community of water users." "The guiding principles of mutual companies," he concluded, "were, first, that rights to land and water should be inseparable, and, second, that water should be apportioned according to need rather than date of settlement, distance from a water source, or even variations in streamflow." MacDonnell (1999, 28) describes ditch companies as "cooperative capitalism."

11. Mormon irrigation communities most closely approximate the communal qualities characteristic of acequias. See Dunbar 1983 and Maass and Anderson 1978.

12. The roles, relationships, and identities tied to the collective provision and use of water are goods in and of themselves; moreover, they are goods participants can and do value from time to time as strongly as water's narrowly liquid properties. See Arnold 2001.

13. Frames are interpretive devices, which, according to Goffman (1974, 21), simplify and condense, permitting individuals "to locate, perceive, identify, and label" events. As applied to collective behavior, social movements in particular, see Snow and Benford 1988 and Snow et al. 1986.

14. By some accounts, more than nine thousand wells were in operation, using upwards of 638,000 acre-feet of groundwater a year (Ogburn 1996, 19).

15. See Curtis 1997, 3. Conservation districts are political subdivisions as well as officially corporate bodies. They may file suits, enter into government contracts, collect fees, and levy ad valorem taxes. County commissioners appoint the members of the conservation district board, provided that each county in the district has at least one member on the board. The conservation district was not the Valley's first explicitly municipal entity for water administration. In 1940, residents along the Conejos River formed the Valley's first water conservancy district. The San Luis Valley Water Conservancy District, encompassing Alamosa, Rio Grande, and Saguache counties, followed in 1949. Conservancy and conservation districts wield similar powers. District judges, however, appoint members to conservancy boards. Residents of the San Luis Valley have formed five of Colorado's forty-six conservancy districts. Together, water conservancy and conservation districts have proven effective platforms for informing and mobilizing residents on a number of water issues.

16. A recent illustration involved Colorado's Statewide Water Supply Initiative (SWSI), a planning mechanism for responding to the state's severe drought of the last several years. Discussions in the San Luis Valley were well attended and environmental, recreational, agricultural, and governmental interests were well represented on the panel. Rick Brown, project manager for SWSI, observing how readily and easily panel members and residents acknowledged diverse concerns and perspectives, commented: "I've seen sensitivity in the meetings in other basins, but not on this level. It's making this process move along quickly." Ray Wright, head of the Rio Grande Water Conservation Board, concurred: "When the valley wants to get something done," he concluded, commenting as much on Valley water history as on the panel discussion, "we get together and get it done." Marcia Darnell, a reporter for The *Denver Post* and resident of Alamosa, put it slightly differently, "Water may be the lifeblood of the San Luis Valley, but unity is its heart." All quoted passages are from Darnell 2003, 37.

17. Six counties comprise the San Luis Valley: Alamosa, Conejos, Costilla, Mineral, Rio Grande, and Saguache. They are among the least populated and least wealthy in Colorado. Census Bureau data from 2000 lists the Valley's population at 46,190 residents, 227 more than listed for 1950. Alamosa is the Valley's most populous community, with 7,960 residents in 2000. Per-capita income for the Valley stands at $14,242, the lowest income region in Colorado. Three counties, Costilla, Conejos, and Saguache, rank among the bottom 10 percent of all counties in the United States. They are just as poor relative to counties in Colorado; among Colorado's 63 counties, Costilla ranks 58th, Conejos 62nd, and Saguache 63rd (San Luis Valley Development Resources Group 2002, B-1, B-8, C-2). For an excellent study on the relationship between poverty and water politics in the

American Southwest, including the upper Rio Grande, see Brown and Ingram 1987.

18. American Water Development, Inc., v. City of Alamosa (Water Div. 3) 874 P2d 352 (1994). For information on the BBC's "Water Wars" documentary, see http://www.nefej.org/av/_tve_vhs/water_wars.htm or http://ftvdb.bfi.org.uk/sift/title/457827.

19. Take, for example, the fight between residents of Costilla County and the Texas-based San Marco Pipeline Company, concluded just prior to the AWDI affair. San Marco sought large amounts of water from the Culebra Creek watershed (approximately 15,000 acre-feet), to slurry coal to power plants in Texas. Foreshadowing the Valley-wide reactions to AWDI, residents of Costilla County fought these plans in court and in the state legislature. They formed the Costilla County Conservancy District to focus their efforts and to cover legal expenses, largely through property tax increases (personal communication with Bob Green, official with the Costilla County Conservancy District, August 31, 2004).

20. The writer Sam Bingham, who witnessed these developments as part of a year-long residency in the Valley, concluded: "Locally, at least, the fight [against AWDI] had made all of them appreciate the vitality and diversity of their desert. They had thought beyond packinghouse and sale barn and the weather, thought about flights of cranes, the racket of migration through the marshes, thought about the fate of 'nonbeneficial' plants and the benefits wrought by people they hadn't thought about before, thought of their valley as a whole piece" (Bingham 1996, 171–172).

21. The notorious "rape" of the Owens Valley is well documented, featuring intrigue and subterfuge by the city of Los Angeles in its acquisition of Owens River water rights in the early decades of the twentieth century (Walton 1992). CSLVW took full advantage of the parallel, arranging for current residents of the Owens Valley to come to the San Luis Valley and recount their valley's fate (Canaly 1997, 13). In short, as L.A. prospered, the Owens Valley spiraled downward, becoming an economically distressed region, bereft of much of its natural vegetation and irrigated agriculture, and home to choking dust storms.

Crowley County, Colorado, east of the San Luis Valley, lost its water to the very cities AWDI hoped to engage (Denver, Pueblo, and Colorado Springs), making it an especially telling comparison. Over three decades, Crowley County's irrigated acreage shrunk by 90 percent, as Front Range cities slowly acquired Crowley County water rights one by one. Although not as well known as the Owens Valley case, Crowley County was not without its own degree of intrigue. As one San Luis Valley newspaper pointed out to its readers, a lawyer purportedly acting as an advocate for Crowley County farmers "proved afterward to be working for metropolitan groups seeking [Crowley County] water" (Hill, 1991, 4). Bare fields and fences buried in wind-blown sand now dot the landscape. Lacking an adequate tax base, the county struggles to meet even basic services. Reflecting on the lessons of Crowley County, Weber concluded: "Now with the drying of the lands, the area's reason for being, its history, and its culture lose their meaning. Metaphorically, the people of the area lose their psychological and cultural 'roots.' . . . For some water exporting areas, it is difficult to imagine a future beyond the present generation (Weber 1990, 13–15).

22. The Alamosa *Valley Courier* relayed the 1990 comments of AWDI's chairman, Lawrence Fox, confirming residents' worst fears: "the 40,000 people in the valley could find someplace else to live" (Hill 1991, 4).

23. Opposition to treating water as a commodity is most clear in the Valley's acequias. Black Mountain Gold, a Texas mining corporation, offered the San Luis People's Ditch $50,000 for temporary water, which they rejected. According to Joseph Gallegos, former majordomo for the San Luis People's Ditch, "The water belongs to the community. It is not for sale" (in Hicks and Pena 2003, 169).

24. Approximately one month before the start of the trial on AWDI's petition, Richard Lamm, already upset over AWDI's litigation costs, publicly invited Valley residents to enter into negotiations and, in effect, make an offer. Portraying water exportation as inevitable—"Nobody wants their water taken. But what are the alternatives?"—Lamm added, "Instead of enriching already rich lawyers, we'd be more than happy to take that money and give it to the valley." John Hill, editor of the *Valley Courier*, angrily dismissed Lamm's invitation, declaring in the process: "This is more than a question of do the citizens of the San Luis Valley want to prevent what happened in the Owens Valley in California and what happened in Crowley County. This is a question of whether SLV residents . . . believe, as former Gov. Lamm implies, that the distribution of resources is the function of the wealthy and that the resources of the state of Colorado do not belong to the people of Colorado. . . . Who controls the wealth?" As to Lamm's assertion that no other alternative to exportation exists, Hill replied: "The alternative is conservation. The alternative is equal distribution for the benefit of all and for the greatest good by agencies who do not have a private agenda" (Hill 1991, 4).

25. Yale University's involvement was not made public until 2002. Outrage over Yale's "secret investment" pressured Yale to contribute substantial sums of money to the Great Sand Dunes National Park proposal (Hunter 2002, B-01; Purdy 2002, B-07; Smith 2002).

26. *Denver Post,* June 3, 1995, B-1. In a later interview with the *Denver Post* (April 7, 1996, B-01), Boyce stated, "No matter how many people want to stick their heads in the sand, the fact that the Front Range needs new and renewable water sources is quite clear. I can demonstrate that we can have a water project without doing environmental damage and economic damage." The $4,000–$7,000 an acre-foot price cited by the *Denver Post* falls within the range characteristic of sales of water in Colorado between 1990 and 2000. According to sales and figures published in both the *Water Strategist* and *Water Intelligence Monthly,* an acre-foot of Colorado water averaged $2,767 between 1990 and 2000. A substantial number of sales, however, topped the $7,000 figure. See Adams, Crews, and Cummings 2004. Boyce may have well expected even greater returns. During the earlier AWDI episode, Bingham (1996, 86) listed the prevailing rates for water as somewhere between $6000 and $10,000 an acre-foot. Steven Shupe (1992, 8), covering AWDI for the *Water Strategist,* concluded that the

200,000 acre-feet of water Strong wanted to sell "could serve one million people and, in the Denver area, could command a price of over $1 billion."

27. *Denver Post,* June 3, 1995, B-1; April 7, 1996, B-01; Quillen 1998.

28. *Denver Post,* June 3, 1995, B-1. While defending his water development proposal in a public meeting, Boyce stated: "Now, I know, people will say, 'Hey, that's makin' money off the water.' Maybe that's going to offend a lot of people's sensibility," an indication in its own way of the moral economy of water. Although aware of residents' views, Boyce candidly admitted his intention of not honoring them. On the prospect of making what he called "a *lot* of money," Boyce replied, "That never offended my sensibility." He ended on a note of ridicule: "A lot of these people who are offended about somebody making money are the same people who aren't offended when it's time to get a grant from the taxpayer. Maybe they'd open their eyes and be reasonable, too." Persons in the audience responded, describing his proposal as "another scam," a "sellout" (Bingham 1996, 233–234).

29. The *Denver Post,* October 20, 1996, 5B. Supporters included an extremely diverse set of groups and interests, among them, The Wilderness Society, San Luis Valley Board of Realtors, SLV Cattlemen's Association, Trout Unlimited, HMO Health Plans, Colorado Insurance Associates, and the Rocky Mountain chapter of the Sierra Club.

30. *Denver Post,* "Water funds top $1 million," October 3, 1998, B-08; Quillen 1998, 16.

31. Initial installation of the water meters would have cost approximately $3 million (Lofholm 1998, A-01).

32. Shriver 1998, H-01; Lofholm 1998, A-01; *Rocky Mountain News,* "'No' On Water Initiatives," October 3, 1998, 63A.

33. As state legislator Carl Miller put it before a rally at the state capitol, "Fifteen and 16 are not about helping our schools. Fifteen and 16 are not about keeping our water in Colorado. Let me tell you what it is about. It's about greed. It's about selling our water to California" (Lane 1998, A-11). Boyce denied that he intended to bankrupt the Rio Grande Water Conservation District. Even so, he did not hesitate to say for the record that, "if that was the result of this law, I wouldn't shed a tear" (Quillen 1998).

34. Between 1993 and 1998, Valley residents wrestled with a suit alleging that its long-standing House District (no. 60) denied Hispanics an equal and fair chance of electing Hispanic candidates, a violation of the Voting Rights Act. The three plaintiffs who initiated the suit wanted the legislature to reapportion the Valley. Their goal was formation of a majority Hispanic district, a goal that would most likely require including areas outside the Valley. Valley residents, on the other hand, drawing attention to the fact that one plaintiff once lobbied for AWDI, saw the suit as a deliberate maneuver to fracture the Valley's famous political unity, making it easier for outside interests to obtain Valley groundwater rights. For Castelar Garcia, who formed Citizens for Solidarity in the San Luis Valley to fight the suit, "This was never . . . a Hispanic vs. Anglo issue. This has been an

issue over our water and those people who want our water," a powerful illustration of water's status as a social good and interpretive frame (Foster 1998, 8A). Opposition to the redistricting proposal relied on a coalition of more than sixty organizations (including conservancy districts, school boards, town councils, and county commissions) and on pointing out the injustice of depriving a valley of its water-related unity and identity. As Valley District Attorney Bob Pastore put it, "We would become another Owens Valley. It could happen so quickly here" (Kowalski 1997, A-01).

35. Editorial, *Rocky Mountain News,* October 24, 1998, 64A.

36. Editorial, "'No' On Water Initiatives," *Rocky Mountain News,* October 3, 1998, 63A.

37. More than 75 percent of the votes cast on amendments 15 and 16 were "no" votes. In Alamosa County, home to Citizens for Colorado's Water, residents cast a total of 5,004 votes; 4,914 were "no" votes (Colorado Election Results, www .rmfc.org/elect98.html#bal; *Colorado Central Magazine,* "Our Spin on the 1998 Election," December 1998, 6).

38. PR Newswire, January 30, 2002.

39. For Valley residents, the injustice of AWDI and Stockman's Water, had they succeeded, would have been the transfer of the effective control over Valley affairs from residents to individuals and entities outside the Valley and arbitrarily privileged by their greater wealth.

40. Chronis 1998, B-05. Compare to the statements of Hill (1991, 4), cited above (n. 24).

41. With respect to complex goods, justice prevails when any one of sphere of meaning or principle of distribution is limited in the interest of preserving the integrity of the rest. Compare to Walzer's principle of complex equality (1983, 17–21).

42. Boyce's efforts to finesse Valley critics are especially illustrative and ironic. Designed to, as he put it, "level the playing field," amendments 15 and 16 actually provided the very forum for critical evaluation he had hoped to avoid. Surely to Boyce's surprise, statewide debates over amendments 15 and 16 only produced even more intense and widespread condemnations of their injustice and unfairness.

43. http://www.leg.state.co.us/2003a/inetcbill.nsf/billcontainers/8733591FFF10 E2AD87256CD20075A5DD/$FILE/HJR1019.enr

44. This moral-economic interpretation of the concept of sustainability includes a number of the features Hempel (1999, 48) associates with "the sustainable communities movement." According to Hempel (1999, 48), "A sustainable community is one in which economic vitality, ecological integrity, civic democracy, and social well-being are linked in complementary fashion, thereby fostering a high quality of life and a strong sense of reciprocal obligation among its members." Unlike the example of the San Luis Valley, however, the sustainable-communities movement is much more urban in its origin and focus. I wish to

express my appreciation to the anonymous reviewer who drew my attention to the sustainable-communities movement and literature.

References

Abbott, Karen. 1998. "Battle Over Water Goes to Voters." *Rocky Mountain News,* October 24, 7A.

Adams, Jennifer, Dotti Crews, and Ronald Cummings. 2004. "The Sale and Leasing of Water Rights in Western States: An Update to Mid-2003." North Georgia Water Planning and Policy Center, water policy working paper no. 2004-004, http://www.h2opolicycenter.org/pdf_documents/water_workingpapers/2004-004.pdf.

Arnold, Thomas Clay. 1996. "Theory, History, and the Western Waterscape: The Market Culture Thesis." *Journal of the Southwest* 38 (Summer):215–240.

Arnold, Thomas Clay. 2001. "Rethinking Moral Economy." *American Political Science Review* 95 (March):85–95.

Bates, Sarah, David Getches, Lawrence MacDonnell, and Charles Wilkinson. 1993. *Searching Out the Headwaters: Change and Rediscovery in Western Water Policy.* Washington, DC: Island Press.

Bingham, Sam. 1996. *The Last Ranch: A Colorado Community and the Coming Desert.* New York: Pantheon.

Blatter, Joachim, and Helen Ingram, eds. 2001. *Reflections on Water: New Approaches to Transboundary Conflicts and Cooperation.* Cambridge, MA: MIT Press.

Blumer, Herbert. 1969. *Symbolic Interactionism.* Berkeley: University of California Press.

Brown, F. Lee, and Helen Ingram. 1987. *Water and Poverty in the Southwest.* Tucson: University of Arizona Press.

Canaly, Christine. 1997. "The Unifying Force of Water." *Rio Grande Water Conservation District Newsletter, Thirtieth Anniversary Commemorative Edition* (September), 13.

Chronis, Peter G. 1998. "Salazar Dives into Water Issues." *Denver Post,* August 13, 2nd edition, B-05.

Corbridge, James N., and Teresa Rice, eds. 1999. *Vranesh's Colorado Water Law.* Niwot, CO: University Press of Colorado.

Curtis, Ralph. 1997. "Formation of the District." *Rio Grande Water Conservation District Newsletter, Thirtieth Anniversary Commemorative Edition* (September), 3.

Darnell, Marcia. 2003. "SWSI Looks at Future of San Luis Valley Water." *Colorado Central Magazine* (December), 37.

Dunbar, Robert G. 1983. *Forging New Rights in Western Waters.* Lincoln: University of Nebraska Press.

Feldman, David Lewis. 1991. *Water Resources Management: In Search of an Environmental Ethic.* Baltimore: Johns Hopkins University Press.

Foster, Dick. 1998. "New District 60 Map Disputed: Critics Say It's about Water, Not Ethnicity." *Rocky Mountain News,* February 3, 8A.

Gascoyne, Stephen. 1991. "The Grit of a Colorado Water War." *Christian Science Monitor,* October 30, 8.

Gillroy, John Martin, and Joe Bowersox. 2002. *The Moral Austerity of Environmental Decision Making: Sustainability, Democracy, and Normative Argument in Policy and Law.* Durham, NC: Duke University Press.

Goffman, Erving. 1974. *Frame Analysis: An Essay on the Organization of Experience.* New York: Harper.

Halaas, David F. 1999. Foreword to *The San Luis Valley: Land of the Six-Armed Cross,* by Virginia McConnell Simmons. Niwot, CO: University Press of Colorado.

Heide, Ruth. 2005. "High Court Justice Addresses Water Challenges, Issues." *Valley Courier (Alamosa),* June 7, 1.

Hempel, Lamont C. 1999. "Conceptual and Analytical Challenges in Building Sustainable Communities." In Daniel Mazmanian and Michael E. Kraft, eds., *Toward Sustainable Communities: Transition and Transformations in Environmental Policy.* Cambridge, MA: MIT Press, 43–74.

Hicks, Gregory, and Devon Pena. 2003. "Community *Acequias* in Colorado's Rio Culebra Watershed: A Customary Commons in the Domain of Prior Appropriation." *Colorado Law Review* 74:101–200.

Hill, John. 1991. "Lamm off Target in SLV Assessment." *Valley Courier (Alamosa),* August 20, 4.

Hughes, Jim. 1998. "San Luis Valley Preparing for Battle." *Denver Post,* October 4, H-12.

Hunter, Mark. 2002. "Yale Helped Fund Plan to Sell San Luis Water." *Denver Post,* January 24, B-01.

Kowalski, Robert. 1997. "Redistrict Bid Splits San Luis Valley: Rift Centers on Race, Water." *Denver Post,* April 20, A-01.

Lane, George. 1998. "Farmers Blast Amendments." *Denver Post,* October 20, A-11.

Lloyd, Jillian. 2000. "Where Ranchers Pine for a National Park." *Christian Science Monitor,* March 22, 2.

Lofholm, Nancy. 1998. "Colorado's Water." *Denver Post,* October 4, A-01.

Maass, Arthur, and Raymond Anderson. 1978. *. . . and the Desert Shall Rejoice: Conflict, Growth, and Justice in Arid Environments.* Cambridge, MA: MIT Press.

MacDonnell, Lawrence J. 1999. *From Reclamation to Sustainability: Water, Agriculture, and the Environment in the American West.* Niwot, CO: University Press of Colorado.

McAvoy, Tom. 2002. "Rural Coalition Unveils 'Water Principles.'" *Pueblo Chieftain,* October 25.

Noreen, Barry. 1994. "Rural Area Beats Back Water Diversion Plan." *High Country News,* vol. 26, no. 10 (May 30).

Ogburn, Robert W. 1996. "A History of the Development of San Luis Valley Water." *San Luis Valley Historian* 28:5–40.

Paddock, William A. 2001. "The Rio Grande Compact of 1938." *Water Law Review* 5 (Fall):1–45.

Pisani, Donald. 1992. *To Reclaim a Divided West: Water, Law, and Public Policy, 1848–1902.* Albuquerque, NM: University of New Mexico Press.

Purdy, Penelope. 2002. "Yale, Water, and Politics." *Denver Post,* February 5, B-07.

Quillen, Ed. 1998. "San Luis Valley Water Goes Statewide." *Colorado Central Magazine,* November, 16.

Rivera, Jose A. 1998. *Acequia Culture: Water, Land, and Community in the Southwest.* Albuquerque, NM: University of New Mexico Press.

Sanko, John. 1998. "Coalition Attacks Southern Colorado Water Issues." *Rocky Mountain News,* October 15, 36A.

San Luis Valley Development Resources Group. 2002. *2002 Comprehensive Economic Development Strategy.* Alamosa, CO.

Saunders, John. 1991. "SLV Residents Fear Same Results as Owens Valley, Crowley County Dust Bowls." *Valley Courier (Alamosa),* October 24.

Sherow, James. 1990. *Watering the Valley: Development along the High Plains Arkansas River, 1870–1950.* Lawrence: University Press of Kansas.

Shriver, Karla. 1998. "The San Luis Valley Water War." *Denver Post,* August 30, H-01.

Shupe, Steven. 1992. "A Tale of Local Concerns: San Luis Valley Fights Interbasin Transfer." *Water Strategist* 5 (January):8–9, 15–16.

Simmons, Virginia McConnell. 1999. *The San Luis Valley: Land of the Six-Armed Cross.* Niwot, CO: University Press of Colorado.

Smith, Erin. 2002. "Allard Thrilled with Results of Yale-Baca Ranch Talks." *Pueblo Chieftain,* January 27.

Snow, David, and Robert Benford. 1992. "Master Frames and Cycles of Protest." In Aldon Morris and Carol McClurg Mueller, eds., *Frontiers in Social Movement Theory.* New Haven: Yale University Press, pp. 133–155.

Snow, David, E. Burke Rochford Jr., Steven K. Worden, and Robert Benford. 1986. "Frame Alignment Process, Micromobilization, and Movement Participation." *American Sociological Review* 10:367–380.

Walton, John. 1992. *Western Times and Water Wars: State, Culture, and Rebellion in California.* Berkeley: University of California Press.

Walzer, Michael. 1983. *Spheres of Justice: A Defense of Pluralism and Equality.* New York: Basic Books.

WCED (World Commission on Environment and Development). 1987. *Our Common Future.* New York: Oxford University Press.

Weber, Kenneth R. 1990. "Effects of Water Transfers on Rural Areas: A Response to Shupe, Weatherford, and Checchio." *Natural Resources Journal* 30:13–15.

3

Ethical Issues in Storm Water Policy Implementation: Disparities in Financial Burdens and Overall Benefits

Sheldon Kamieniecki and Amy Below

Once sparsely populated, Southern California has become rapidly urbanized over time. As land surface has become sealed with pavement and buildings, larger amounts of pollutants are washed away by storm water. Unlike many other locations, the polluted storm water is not treated and washes directly into receiving waters and the ocean, consequently degrading the marine environment. Contaminated water contains bacteria and pathogens and also harms swimmers and surfers. According to Dwight et al. (2005), tens of thousands of individuals annually contract gastrointestinal illnesses, stomach ailments, and respiratory, eye, and ear infections. Those who become ill accrue about $3 million annually in health-related costs. In addition, increased levels of storm water flow lead to greater flooding and prevent replenishment of groundwater aquifers. Legal actions taken by environmental groups under provisions of the federal Clean Water Act and the California Porter-Cologne Water Quality Control Act have forced federal, state, and local governments to try to reduce the level of pollution in storm water runoff.

Treatment plants like those used for sewage are not practical for handling storm water in municipalities that currently have separate wastewater and storm water collection systems. Storm water flows are quite irregular, and the capacity necessary to treat peak flows is extremely expensive (Gordon et al. 2002). Instead, it is more likely that management will consist of a broad system of "best management practices" (BMPs) in combination with aggressive source control (Devinny et al. 2005). Such a system is currently being developed by the Los Angeles Regional Water Quality Control Board (LARWQCB) for a large section of Southern California. The LARWQCB is a state regional agency that manages ground and surface water quality in Southern California, specifically the coastal watersheds of Los Angeles and Ventura counties and small portions of Kern and Santa Barbara counties. Under the current plan, the costs and benefits of controlling pollution in storm water runoff will be unequally distributed throughout the area. This raises significant equity and policy issues regarding the long-term effort to control water pollution in the region, state, and other areas of the country.

In this chapter, we analyze the extent to which storm water runoff policy in Southern California is equitable and ethical. We begin by presenting a brief overview of the storm water runoff problem. Next, we examine economic, race, and ethnic divisions in Los Angeles County and discuss the equity issues involving municipalities close to the ocean, those that are located in the inner city areas of the County, and those that are upstream in rural areas. We point out how coastal communities benefit the most from efforts to control storm water pollution (e.g., from increased property values and income from local visitors and tourists), while inland-urban and rural communities must sacrifice the most for storm water abatement. Inland residents tend to spend less time at the beach and take less advantage of recreational activities related to the ocean than coastal residents. Those who pursue a rural lifestyle and engage in agricultural and other activities such as hiking, camping, and horseback riding will have to pay for storm water quality control but will also not benefit as much as inhabitants of beach communities. Such patterns of inequity raise environmental justice concerns largely ignored by previous research as well as environmental groups and policymakers. The chapter identifies costs associated with pollution control and abatement. Financial and other benefits, such as groundwater replenishment, more appealing beach environments, improved public health, and the creation of additional urban green space, are also noted. The chapter presents an in-depth, theoretical analysis of the different ethical concerns that should be taken into account by government in environmental policymaking. By paying closer attention to ethical issues, decision makers will be more likely to choose equitable approaches to environmental protection. Exactly how this can be done within present storm water runoff policy in Los Angeles County is examined at the end of the chapter. Clearly, water is a scarce resource in the Southwest. However, selected management techniques for storm water runoff must not only result in improved water quality and water supply, they should also take into account critical equity issues as well.

The Storm Water Problem

Most urban runoff is generated during rainfall events and can properly be termed storm water.[1] This flow is extremely irregular, especially in Southern California, where most days are dry, and measurable rain occurs on average of only 32 days per year. Total rainfall in the area is modest, averaging about 16 inches per year. A large storm in this area might drop as much as three inches of rainfall in 24 hours, but this is still much less intense than typical rainfall events in other regions of the country such as the Northeast, Southeast, and Midwest.

Nevertheless, high flows and flooding do occur in Southern California because of the topography. Water from large watersheds drains into local rivers and streams. Many slopes are steep, and rainfall is, thus, rapidly collected and concentrated.

Water also enters the storm drains from nonrainfall sources. Irrigation systems for watering gardens and lawns, automobile washing, and the hosing down of sidewalks and driveways generate smaller streams sometimes referred to as "nuisance" flows. This water flows into the storm drain system all year and, with residual stream flows (and in certain areas, recycled wastewater), constitute dry weather flow. The terms *storm water* and *runoff* are often used interchangeably. However, there is a major difference between the two: storm water arrives suddenly in large amounts while runoff flows are considerably smaller and occur all year.

Urbanization of the landscape dramatically changes the amount and composition of runoff. Because less water infiltrates (percolates) into soil, the overall amount of runoff is increased. Since the water runs off pavement more rapidly, it is concentrated to make peak flows higher. Recharge of groundwater is substantially reduced, and the shallow groundwater that feeds some streams dries up. As a consequence, surface flows decrease in some areas. Surface flows may increase during dry weather in other communities because of nuisance flows from over-irrigation, automobile washing, and other sources. In general, the storage and buffering effects of soils and groundwater reservoirs are reduced. Runoff flowing through vegetation, or entering and leaving shallow groundwater, is subject to the effects of filtration and biodegradation, which has a significant purifying effect. Water runoff from pavement is not cleaned and is contaminated by whatever dirt and pollutants are on the pavement.

Runoff pollutants tend to vary a great deal and are different from those in sewage. Bacteria and pathogens from animal and human waste are present but in much smaller concentrations than found in sewage.[2] The same is true for nutrients such as phosphorous and nitrogen. There are more petroleum hydrocarbons, dust, sediments, and settled air pollutants in runoff, but total organic content is lower than in wastewater.

The pollutant load of storm water varies considerably with location. Storm water contains pollutants that wash off rooftops, parking lots, industrial facilities, and streets. Pollutants may also be discharged illegally, such as when industries discharge toxic pollutants. In June 2005, the state attorney general and three county prosecutors sued the automobile-parts chain AutoZone, accusing it of illegally dumping hazardous waste in storm drains (*Los Angeles Times* 2005).

Water flowing in the streets picks up trash (cigarette butts are a major problem), dust, dirt, and other materials that have been deposited on the pavement. The dust includes fine particles of rubber from tire wear, settled air pollutants, trace metals from brake pads and other mechanical sources, and pet feces. Automobiles, buses, and trucks drip motor oil onto the pavement and the early flows of fall (when the rainy season begins) tend to carry a petroleum sheen.

Storm water and runoff come from a variety of sources, each one producing a different combination of pollutants. Streets, especially those in dense commercial areas, are the most difficult source of urban runoff to control. In addition to their huge number in Southern California, they tend to vary substantially in size, location, and pollutant sources. Large-scale manufacturing and commercial activities, such as oil refining, tend to denigrate storm water quality. Rain falling on uncovered machinery, materials, and contaminated surfaces picks up pollutants and carries them into streets and storm drains. Similarly, construction sites also produce pollutants that are carried away by runoff. Roof runoff from commercial facilities can contain air pollutant dust, bird droppings, hydrocarbons from roof tar, and some trace metals from rooftop machinery. Nearly all parking lots are designed for rapid drainage to the street or storm drain. Parking lots usually contain litter, heavy metals from motor vehicles and road wear, and oil leaking from motor vehicles. Finally, single-family homes produce considerable roof runoff containing dust, bird feces, and settled air pollutants. Runoff from lawns and gardens often include pesticides, herbicides, insecticides, and large amounts of fertilizer. Some homeowners disobey the law and dump motor oil, paint, and other toxic chemicals into storm drains.

Protection Measures and Their Costs

Storm water quality protection measures fall into three general categories. Infiltration allows percolation of the water into the ground, relying on the soil to remove pollutants from the replenishing groundwater and eliminating the discharge to runoff. Source control measures prevent the release of pollutants, so that the water is never contaminated. Treatment systems remove the pollutants from the storm water before it reaches the ocean. Clearly, any solutions requiring storm water runoff be diverted to existing sewage treatment plants will be extremely expensive. Besides expanding capacity of existing plants, it will be necessary to provide additional technology to treat a large and complex variety of runoff pollutants not found in sewage. This explains why the LARWQCB has adopted a system of less costly BMPs to improve the quality of storm water and runoff.

Thus far, the LARWQCB, the State Water Quality Control Board, and the Environmental Protection Agency (EPA) have ignored protests from mainly inland and rural communities over the costs of controlling storm water and runoff pollution. Members of the LARWQCB have been ordered by the EPA and the courts to develop and implement a plan in accordance with the Clean Water Act that will curtail pollution in storm water runoff and protect the ocean. The Coalition for Practical Regulation (CPR), an ad hoc group of about forty communities in Los Angeles County, has argued that the new plan as it is currently formulated will cost over $50 billion and is unnecessary because most cities are already addressing storm water issues (Water Environment Federation 2004). Developers, businesses, the Los Angeles County Board of Supervisors, and local community groups such as Mothers of East Los Angeles (a predominantly Hispanic organization) and Concerned Citizens of South Central Los Angeles (a predominantly African-American organization) have supported legal and other actions taken by the CPR. The LARWQCB and environmental groups such as the Natural Resources Defense Council, Heal the Bay, Tree People, the Surfrider Foundation, and the Santa Monica Baykeeper argue that this figure is a gross exaggeration and, without tougher measures in place, the quality of storm water runoff and the ocean water will continue to decline significantly (Beckman et al. 2002). Academic researchers also agree that the $50 billion figure is very far above the true cost for required storm water cleanup (Devinny et al. 2005).[3] Ignored in this debate thus far are the substantial economic, racial, and ethnic disparities that exist between those communities that will have to shoulder the financial burden of storm water cleanup and those communities that will most likely profit from this effort.

Regional, Economic, Race, and Ethnic Divisions

A statewide survey directed by Mark Baldassare for the Public Policy Institute of California (2003) shows that a majority of Californians (52 percent) consider ocean and beach pollution a serious problem, and more than seven out of ten state residents are concerned about the decline in marine mammals along the coast. Overall, 69 percent of Californians said the condition of the coastline is very important to their quality of life, and 61 percent consider it very important to the state's economy. More than 70 percent favor reducing pollution from storm drains or sewage plants, even if the cost of abatement leads to higher utility bills. Almost 75 percent of respondents reported visiting the coast at least several times a year. According to the survey, 43 percent swam in the ocean, 17 percent went fishing, 14 percent went sailing or kayaking, and 10 percent went surfing.

Despite shared concern for their 1,100 mile coastline, the survey indicates that there are regional differences in Californians' attitudes regarding coastal and environmental issues. In particular, Southern Californians place greater importance on the shoreline, are more concerned about worsening coastal conditions, and are more personally connected to the ocean and beaches than those who live in other regions of the state (e.g., Northern California and inland areas). More Southern Californians (74 percent) than residents of Northern California and inland areas (67 percent and 62 percent, respectively) feel the condition of the coastline is very important to their quality of life. Southern California residents are also more likely than residents in other regions to say the ocean and beaches are very important to the economy. Many more people living in Southern California than in other regions of the state think ocean and beach pollution is a serious problem. Fifty-seven percent of Southern California inhabitants believe the condition of the ocean has grown worse in the past twenty years, while 44 percent of Northern Californians and 50 percent of inland residents share this view. A higher percentage of those living in Southern California surf and swim in the ocean than people living in other areas of the state.

The survey reveals interesting ethnic differences as well. Hispanics are more concerned than non-Hispanic whites about many of the environmental problems affecting the coastline. For instance, they are more likely to view as serious problems ocean and beach pollution (66 percent to 49 percent), the contamination of seafood (64 percent to 46 percent), declining numbers of sea mammals (54 percent to 40 percent), and public access to the coast (27 percent to 17 percent). Willingness to fund storm water pollution abatement, particularly in comparison to white residents, was not reported by the Public Policy Institute of California. Unfortunately, the Institute did not report any survey findings involving economic or other racial and ethnic groups.

Previous research shows there is a negative relationship between willingness to pay and distance from the body of water one seeks to preserve (Sutherland and Walsh 1985). Persons living farther from the ocean are less likely to visit the beach. Moreover, as visit rates decline with increased distance from the site, so, too, does willingness to pay for nonuses of the body of water (Sutherland and Walsh 1985). With respect to the storm water issue, those living in inland communities may have a lower willingness to pay than those in coastal communities but could be forced to pay for the majority of control measures. Compounding the issue is the likelihood that these inland communities may also have less ability to pay.

Demographic data reveal significant differences between economic and racial and ethnic groups and how close people live to the coast in Southern California. As table 3.1 shows, coastal, rural, and inland-urban communities in Los Angeles County tend to vary considerably across different economic characteristics. Residents who live near the coast tend to have higher median house values and possess higher median household income levels than residents who live in rural and inland-urban municipalities.[4] Coastal residents are also more likely to be employed than those living in rural and inland-urban communities. The data clearly show, therefore, that those who are already better off financially are likely to benefit the most from controlling storm water pollution. Likewise, those who are less wealthy will most likely pay a proportionately greater amount for pollution control and benefit less from the policy.

In addition to the need to protect high property values, Pendleton (2001) estimates that local area residents and tourists who visit Southern California beaches spend over $1.5 billion on food, goods, supplies,

Table 3.1
Economic characteristics of Los Angeles County communities by geographic location

Community	Median value of homes by city, 2000 census	Median household income in cities, 2000	Per capita income in cities, 2000	Unemployment percentage rate by city, November 2004
Coastal average	$538,014	$84,229	$48,341	3.0
Rural average	$196,867	$51,323	$24,908	4.3
Inland/urban average	$232,411	$47,692	$19,752	5.5

Sources: Median and Per Capita Income in Cities and Unincorporated Communities, Los Angeles County, 2000.Los Angeles Almanac. 2005. http://www.los angelesalmanac.com/topics/Employment/em12.htm Labor Force by City and Unincorporated Community, Los Angeles County (data not seasonally adjusted), Recent Preliminary Data—November 2004. Los Angeles Almanac. 2005. http://www.losangelesalmanac.com/topics/Employment/em03.htm
Median Value of Homes by City and Unincorporated Communities, Los Angeles County, 2000 and 1990

and recreational activities. Although the quality of coastal waters has improved somewhat or remained about the same over the last five years, many people in Southern California still perceive the ocean too polluted and avoid bathing in it (Pendleton et al. 2001). This is primarily due to frequent news reports of ocean contamination warnings, partial beach closings, and continued incidents of sickness from swimming in the ocean. Unless more is done to improve ocean quality, coastal communities may suffer a significant decline in commerce and housing values.

Table 3.2 demonstrates how much coastal, rural, and inland-urban communities in Los Angeles County vary in terms of racial and ethnic composition. As the data indicate, high percentages of whites tend to live in coastal communities. In fact, in most coastal communities whites comprise two-thirds or more of the population. On average, just over one-quarter of residents in inland cities are white. In many of these cities the population is less than 10 percent white. Hispanics inhabit inland communities more than any other minority population. Reflecting the diverse nature of Southern California, other ethnic and racial groups are well represented in inland cities as well. Rural communities, in contrast,

Table 3.2
Racial/ethnic characteristics of Los Angeles County communities by geographic location

Community	White	White (not Hispanic or Latino)	Black or African American	American Indian and Alaskan Native	Asian	Native Hawaiian and other Pacific Islander	Some other race	Population of two or more races	Hispanic or Latino
Coastal average	73.30%	67.16%	3.58%	0.38%	12.97%	0.44%	5.33%	4.00%	12.64%
Rural average	67.80%	57.73%	10.55%	0.80%	4.41%	0.74%	11.53%	4.74%	23.93%
Inland/urban average	48.09%	27.16%	6.44%	0.87%	15.49%	0.28%	24.04%	4.65%	48.20%

Source: Racial/Ethnic Composition of Cities by Percentages, Los Angeles County, 2000 Census. Los Angeles Almanac. 2005. http://www.losangelesalmanac.com/topics/Population/po38.htm

have a higher percentage of whites than inland cities, though not as high on average as coastal communities.

Environmental Justice and Pollution Control

Up until the beginning of the 1990s, pollution control was treated primarily as a rather narrow scientific exercise by most planners and policymakers. Although the debate over whether to regulate was a major political issue, once it was decided that government action was necessary legislators and agency officials sought to identify which technical solutions best prevent and abate contamination at the lowest cost. Following the passage of major pollution control legislation, many of the nation's environmental problems were delegated to individuals in the EPA with technical backgrounds to analyze and resolve. For the most part, EPA officials and even environmental groups ignored the impact government policies might have on particular social groups.

Since the early 1990s, the environmental justice literature, along with the environmental justice movement, has revealed that poor and minority communities are disproportionately exposed to various types of pollution and its risks, and that the implementation of government remedies tend to be slow in these communities in comparison to largely white, affluent areas (e.g., Bryant and Mohai 1992; Mohai and Bryant 1992; Bullard 1994a, 1994b; Bryant 1995; Camacho 1998; Simpson 2002). The environmental justice literature has shown, for example, how air and water pollution and toxic waste tend to affect poor and minority communities more than other areas and that government has been slow to abate pollution or remediate contaminated land in these locations. Many poor and minority communities are home to grossly polluting industries, hazardous waste facilities and dumps, and other toxic sites. Among the most prominent incidences of environmental injustice are Love Canal, New York; Warren County, North Carolina; and "Cancer Alley" in Louisiana. Although well intentioned, federal and state legislation has historically ignored this phenomenon and in some cases has exacerbated the problem. Those active in the environmental justice movement have called for changes in government policy in order to alleviate social, geographic, and procedural inequities.

Environmental justice studies also argue that allowing the disadvantaged to be continuously exposed to dangerous contaminants more often than predominantly white and wealthy individuals is not only unethical and immoral, but it also violates central tenets of American democracy and the U.S. Constitution (e.g., Bowman 1997; Kamieniecki and Steckenrider 1997; Schlosberg 1999). Everyone has an equal right to a clean environment and good health, and no single group should be forced to endure extended contact with dangerous pollutants. It is not only the responsibility of government to protect its citizens from air and water pollution and harmful toxic chemicals and waste, but also government must ensure that certain segments of society do not receive less protection than others. The distribution of the costs and benefits of environmental protection should be shared equitably by all Americans. A situation where poor and minority communities must absorb higher costs and obtain fewer benefits of pollution control should not be permitted on ethical, moral, and legal grounds. Yet, this scenario currently exists in Southern California regarding storm water policy.

There are many possible causes of environmental injustice (Foreman 1998). Researchers often point to the lack of financial and political clout of poor and minority groups and, at the same time, the considerable influence affluent, white communities have over government actions (e.g., Schlosberg 1999). Persons in poor, minority communities generally lack the financial resources and well-placed connections necessary to sway government policymaking. They also tend to participate in politics at lower rates than white citizens (Verba and Nie 1972; Conway 2000). Historically, they possess little political leverage and present less political resistance. According to Camacho (1998), in the end the increased risks within such communities are the result of social inequality and imbalances of power. As he correctly observes, "'losers' suffer a cumulative systemic disadvantage and become outsiders to the governmental process" (1998, 18). In effect, certain groups are left out of the process entirely. In some cases, poor and minority citizens may support the siting of polluting facilities in their community for financial gain and thus feel obliged to accept the risks in order to survive economically.

Although the environmental justice literature has made an important contribution by pointing out many inequities in environmental policymaking across the country (e.g., Mohai and Bryant 1992), the literature has also been criticized on a number of grounds. As Bowman (1997), Kamieniecki and Steckenrider (1997), Foreman (1998), and others demonstrate, various studies contain serious methodological flaws and have produced inconsistent findings. Early studies (e.g., analyses of Superfund sites) were anecdotal and thus provided only subjective evidence of environmental injustice. Recent studies, however, are more rigorous and employ sophisticated and complex methods of analysis (Ringquist 2003). In addition, a number of scholars base their conclusions on subjective perceptions of environmental justice and ignore scientific studies of health risks (Foreman 1998). Living in a community with a toxic waste facility does not automatically mean that all residents are suffering health problems.

Another critique of the environmental justice literature is that researchers tend to focus too narrowly on race, omitting other categorical groups (e.g., social class) and other populations (rural residents).[5] For example, persons who have chosen specific lifestyles such as farming may be disproportionately exposed to risks or otherwise adversely affected by policy decisions. Those who work in agriculture are likely to come into contact with toxic chemicals such as fertilizer, insecticides, and herbicides. Active and abandoned mines throughout the Southwest also expose rural residents to dangerous pollutants such as cyanide, arsenic, mercury, sulfuric acid, and heavy metals. The environmental justice literature can benefit from broadening its scope to include other segments of society that have also been exposed disproportionately to pollution for a period of time.

While environmental justice scholars have effectively criticized inequities that exist in pollution control and abatement and, in so doing, provided thoughtful discussions of the unethical and immoral nature of certain government policies, procedures, and practices, they have failed to outline specific solutions in cases where such inequities exist. Just as most environmental policymakers and environmental groups have viewed the abatement of serious pollution problems in strict technical and scientific terms, most environmental justice researchers (and advocates) have studied the costs and benefits of environmental protection in primarily racial terms (e.g., Bullard 1994a). Policymakers, environmentalists, and environmental justice analysts have failed to move beyond their own belief systems and paradigms and offer specific policy approaches that are both effective and equitable in the way they distribute costs and benefits. Instead, policymakers concentrate their efforts on developing cost-effective scientific approaches, environmental groups lobby for significant risk reduction and complete abatement of pollution, and environmental justice advocates demand the protection of minorities, particularly African-Americans, at any cost. Today's environmental problems, such as the pollution of storm water runoff in Southern California, deserve solutions that address and integrate these perspectives. Of course, aiming for such an ambitious outcome will invariably pit equally valid equity concerns and goals against one another, requiring present institutions to adopt more inclusive collaborative decision-making processes that can provide opportunities for debating and deciding on alternative, equitable policies.

Environmental Justice and Storm Water Runoff in Southern California

As this analysis suggests, storm water runoff policy in Southern California overlooks environmental justice and equity issues. A main reason for this is that neither the federal Clean Water Act nor the California Porter-Cologne Water Quality Control Act and their amendments require policymakers to consider environmental justice problems in the formulation and implementation of rules, regulations, and programs concerning water quality. Although a citizen suit provision exists in the Clean Water Act (O'Leary 2003), primarily major environmental groups employ it in seeking new regulations and enforcement. For the most part, mainstream environmental organizations have ignored environmental justice issues in their lobbying activities and legal actions. Citizen suit provisions are not included in California environmental laws. As a result, federal and state policies fail to take into account place of residence, economic, racial, and ethnic disparities and are, as a consequence, unethical.

Even if existing legislation was revised accordingly, institutional barriers to developing equitable and ethical environmental policies would still exist. As in most areas of the country, storm water management (as well as many environmental policies more generally) is highly fragmented in Southern California. In addition to a lack of cooperation between levels of federal, state, county, and municipal government, each government agency jealously guards its decision-making authority, with citizens, stakeholders, and officials from other agencies largely confined to commenting on proposed standards and regulations. Among other things, this tends to isolate the voice of minorities and the poor in the decision-making process and makes it difficult for these citizens to influence policy outcomes. A far better approach would be to focus on all sources of a pollutant within a watershed as a whole, rather than on types of sources (that is, point and nonpoint) or on the arbitrary political boundaries of states, counties, and municipalities. In addition, a new institutional structure is needed where the decision-making process is collaborative and involves face-to-face negotiations among a variety of stakeholders with relatively consensual decision rules. Rather than having each agency single-mindedly pursue its legal mandate within a particular jurisdiction, the new institutional structure would seek win-win solutions to an inter-related set of social, economic, and environmental issues confronting the watershed as a whole. The result would be a collaborative process involving extensive negotiations over a period of time, instead of a series of standardized rule-making decisions whereby an agency proposes a rule, receives comments, revises the rule, and then awaits litigation by one or more unsatisfied stakeholders. This is exactly the approach the LARWQCB has taken thus far. This new, collaborative approach represents a significant shift from a top-down, agency-dominated approach with some provision for citizen comment to a much more collaborative, bottom-up strategy involving negotiations and problem solving among a variety of governmental and nongovernmental stakeholders. While the adoption of such a collaborative process cannot automatically guarantee the selection of equitable and, for that matter, effective environmental policies, it stands a greater chance of addressing environmental justice and ethical concerns in policymaking efforts.

Admittedly, the issue of storm water runoff in Southern California does not directly resemble traditional environmental justice issues. Storm water and runoff are not point sources like polluting factories or toxic dump sites. Nevertheless, as already argued, attempts to address storm water issues most certainly include race and class considerations and un-equal distributions of wealth and power across and among regions and places. Where environmental justice potentially comes into play is in the determination of the costs and benefits of controlling storm water pollution in the region. Based on present policy, costs will likely be dispropor-tionately paid by inland racial and ethnic minority-urban communities as well as rural residents while the benefits of a cleaner ocean will be enjoyed primarily by the more advantaged coastal communities.

Furthermore, as with other environmental justice problems, the more affluent coastal communities are likely to influence government decision making while their counterparts will find themselves subject to policy mandates on which they have had little input. Although the LARWQCB has held many public meetings, inland and rural municipalities say the Board has not consulted widely and has not been accessible enough to the public at large (Forester 2004). This also may explain why certain environmental justice issues have not been addressed in present storm water pollution control policy.

Ethical Considerations in Storm Water Policy Design

If environmental justice issues are to be meaningfully addressed in the formulation and implementation of storm water pollution control policy, policymakers will have to alter their mind-set and formally consider the role of ethics in policy design. Currently, policy design involves, among other things, a highly calculative process comparing costs and benefits. If the benefits (e.g., improved air quality) of a policy approach outweigh the costs (e.g., new automobile emissions standards), the policy approach is considered worthwhile and is selected for implementation. The reverse is also true. If costs outweigh benefits, the policy option is considered undesirable and it is not adopted. Similar in concept to cost-benefit analysis, yet less often employed in policy design, is ethic-based evaluation.

Ethical behavior has been discussed since the time of the earliest philosophers. In our era, individuals often consider ethics in their personal lives, but ethical considerations have largely been excluded from the evaluation of group behavior in government, business, and industry. However, considering the inequities prevalent in some environmental policies, as evidenced by findings in the environmental justice literature, ethical considerations should be incorporated into the decision-making process. Evaluating policy options on ethical and moral standards has the potential to recognize and rectify current and future inequities.

One of the most common points of reference in policy evaluation is policy outcome. A policy is evaluated by assessing its end result. This is known as ends-based or "consequentialist" evaluation. When applying such teleological moral theory, the end result is of higher importance than both motive and the means or rules by which the policy was implemented.

Within consequentialist thought, one of the most debated theories is that of utilitarianism. Within the ethics and philosophy literature, utilitarianism "is the doctrine that the morality of an action depends solely on its consequences" and that "if the consequences are good the act is right and if the consequences are bad the act is wrong" (Singer 1977, 67).[6]

The idea of utilitarianism frequently surfaces in studies of environmental policy (as well as in studies of many other public policies), where the aim is the greatest good for the greatest number of people. In this way, "utilitarian theories are teleological theories with a welfarist theory of value" (Brink 1986, 421). Human welfare or happiness is of highest value, specifically the welfare of the many.

In addition to the end result, according to many utilitarian thinkers, it is also important to consider the range of possible alternative actions. Here the right choice is made when the best alternative is chosen, which is the action leading to the best consequences. Thus, according to the same logic, a decision is wrong if another alternative would have produced better consequences even though the chosen route did produce some good. It is not enough to simply increase the amount of happiness in the world; one must choose the action that increases happiness more than any other option would. Thus, a decision maker should look for the alternative with the *best* consequences. This would be the most ethical decision. In other words, "An action may be productive of benefit and still be wrong, if the agent in the circumstances could have done something else that would have produced a greater benefit" because one is aiming for the greatest good for the most people (Singer 1977, 67).

The significance of determining and evaluating the "rightness" or "wrongness" of consequences brings to the fore the question of which consequences should be judged: the intended, the probable, or the actual. Moore (1947) is a proponent of "actual consequentialism." For him, the actual consequence determines what is right and wrong. Again, it is the end result that matters, nothing else, not even intent. Within this frame of mind, even well-intentioned acts affected by unforeseen occurrences can make a once thought right action wrong (Moore 1947). While one may not be culpable and may be forgiven for this kind of wrongdoing, his/her actions are still misguided because they resulted in a bad outcome. In this case, the ethical value of the decision is decided regardless of and in spite of intention.[7]

A specific type of utilitarianism focusing on such an aspect of ethical evaluation is that of "motive utilitarianism." In this instance the motive behind an action is considered, but utilitarian ideals are still paramount. According to Adams, "Perfect motivation is identified with an all-controlling desire to maximize utility" (1967, 468).[8] Simply put, a motive is good if it produces happiness or pleasure for as many people as possible and bad if it produces pain. Moreover, if adhering to the idea of degrees of goodness, the more good an action produces, the more correct the motive was. Thus, here the ethical value of an action can be assessed by considering motive, but the motive is still ultimately judged by the outcome it produces.

Naturally, such a method of evaluation has its share of problems. One difficulty stems from possible discrepancies between motives and consequences. For example, one may be motivated by selfishness, but the consequences of his/her actions in the end promote general happiness. Hence, it is possible that other motives besides purely utilitarian ones could have the most universal benefit. In such cases, what is right by "motive-utilitarian" standards may not be under "act-utilitarian" standards. The motive does not necessarily have to match the act. This leaves the evaluator with a choice: evaluate the motive, the act, or the end result. Each may produce a different conclusion.

In addition to the complexities already discussed, there are also several general moral objections to the principle of utilitarianism. For instance, it "fails to accommodate the extent of our obligations to others, the existence of moral and political rights, and the demands of distributive justice" (Brink 1986, 418). It also fails to consider "that people lead separate lives, possess different commitments, and pursue different projects and plans" (Brink 1986, 418). Utilitarianism lumps everyone into a homogenous group assuming that their needs are uniform and that their life situations are identical when, in fact, this is most often not the case. Thus, what might benefit or bring happiness to some, and thus be termed right or ethical, may not have the same positive effect on others. Maximizing utility and satisfaction over what are most often in reality heterogeneous populations is a complex and difficult task.

In terms of utilitarianism, and consequentialism in general, the problem then arises that what would be a good consequence for some could be bad for others. Similarly, there could be degrees of good and bad consequences interspersed within the population. In the case of the storm water runoff issue, a specific policy might have positive effects in some communities (e.g., coastal ones) but deleterious effects in others (e.g., inland ones). Southern California is decidedly not homogenous, and each policy option will affect different social groups differently. Thus, it is important, if applying a utilitarian ethic, to be truly utilitarian by considering the needs, desires, and situations of the entire affected population. The greatest good needs to be applied to the greatest number of people possible with the understanding that all individuals should be awarded an equal amount of the "good" regardless of, in this case, geographical location and more especially, social standing or racial and ethnic background. Similarly, all individuals should have to bear an equitable—not necessarily equal—amount of costs. It is possible that the greatest good might require trade-offs—and that a fair burden must be assessed in proportion to one's ability to pay, and in proportion to the benefit he or she receives. This suggests that those who will reap the biggest benefits should pay a larger portion of the costs so that, on balance, the net rewards and costs are fairly, and ethically, dispersed.

Perhaps even more applicable to the environmental justice and storm water issue is the specific consideration of the relation between ethics and politics. According to Morgenthau, "The aim of moral action is the attainment of the greatest amount of human satisfaction" (1945, 1). However, Morgenthau points out what he sees as a clear distinction between the private and public spheres. Each has its own ethical code, the code in the private sphere being more stringent as actors in the political sphere violate ethical standards they would not when acting on their own. When acting on the behalf of and in the best interest of the common good, they are allowed to, and are almost obliged to, behave in ways that are at times considered less ethical because it is the end result of the common good that is paramount.

Thus, in politics many apply the consequentialist theory; the ends are more important than the means and can even sometimes justify unethical means to attain them. As Morgenthau summarizes, "The ends taint the means employed for its attainment with its own ethical color and thus justifies or condemns that which, considered by itself, would merit the opposite valuation" (1945, 7). Some politicians are willing to forego some ethical behavior if it means that, in the end, the population is better off (or, as they perceive them to be better off).

To be sure, such a justification of unethical behavior is not universally accepted. Beneficial ends cannot always justify unethical means. For example, many argue that it is unethical for one group to suffer in order to protect the welfare of another. Beneficial consequences do not warrant the mistreatment of one group en route to the happiness of another. The fact that certain communities will benefit from storm water controls and the enhanced quality of the ocean water does not justify that other communities must pay most of the costs, even if the perceived benefit is ubiquitous. This logic contrasts with utilitarianism by placing value on the means whereby a policy is implemented. While the end result is still important, the means used to get there should not be discounted. The means to the end should be ethical as well, and that means ethical for the entire population.

Morgenthau (1945) also tackles the issue of political intent. He believes that political actors have a moral obligation to act wisely. Whereas, if acting alone, one's unwise action hurts (hopefully) few people, as a public servant the repercussions of unwise action affect many. Thus, he writes that what is done "with good intentions but unwisely and hence with disastrous results is morally defective, for it violates the ethics of responsibility" (1945, 10). Even the best of intentions can result in unethical ends. Hence, Morgenthau supports the idea that consequences are not the only consideration. The means must be prudent and ethical as well.

Riley (1990) also addresses a criticism of utilitarianism in his discussion of the ethics of democracy. He warns of the unequal distribution of utility and writes that, "Aggregate utility maximization might imply a highly unequal distribution of personal satisfaction such that a privileged elite enjoys great happiness while the majority grovels in great misery" (1990, 336). His critique reiterates the difficulty with evaluating acts on a purely consequentialist and utilitarian basis. It is not enough to look at the overall benefit to society, in this case enhanced water quality. One must also consider how the benefits are dispersed within society and ensure that they are allocated equitably, and thus, ethically.

This leads to the issue of procedural utilitarianism or equality. In terms of democratic decision making, the utilitarian perception of procedural equality implies that everyone is afforded equal opportunity to affect policy decisions as each person's happiness is equally weighted. This is utopian because all citizens are "not treated equally unless they all happen to share identical cardinal comparable personal utility functions" which they do not (Riley 1990, 340). Thus, there is often little actual utilitarian value in procedural utilitarianism. Everyone's opportunity to affect policy outcome is not equally weighted. Some are provided more opportunities, while others are victim to others' decisions. As Morgenthau writes, "The test of a morally good action is the degree to which it is capable of treating others not as a means to the actor's ends but as ends in themselves" (1945, 14). A policymaker engaging in ethical behavior would treat everyone equally and would consider how the consequences affect each segment of the population.

A More Ethical Storm Water Policy

At this juncture the LARWQCB has adopted a number of different BMPs that will help to control pollution in storm water runoff and will improve the quality of receiving waters, including the ocean. Briefly, the current list of BMPs include increased enforcement of litter and dog waste ordinances; chemical use, storage, and spill prevention regulations; improved street sweeping, recycling, and vacant lot clean-up programs; public education; programs to prevent illicit discharges; and storm drain cleaning requirements.[9] Moreover, the Board has adopted numerous rules and regulations governing business activities and new construction and development in the region.

Collectively, these BMPs place a heavy financial burden on private firms and especially numerous inland-urban and rural municipalities. For example, the upgrade of street sweeping in Los Angeles County will require municipalities to purchase new, advanced vacuum-type sweepers to replace those currently in use (Devinny, Stenstrom, and Kamieniecki 2005). Presently, there are about four hundred street-sweeping machines in use. These machines must be replaced once every four years, thus, one hundred machines are replaced each year. A vacuum machine costs about $150,000 in comparison to $75,000 for a current standard machine. Therefore, the additional cost for a higher quality sweeping machine is $75,000 above the cost for the standard machine, or a total of about $7.5 million each year (Devinny, Stenstrom, and Kamieniecki 2005). Maintenance costs for the vacuum-type sweepers are also higher than for the standard machines.

By using the central principles of utilitarianism, and consequentialism in general, along with the philosophy of Morgenthau (1945) as a guide, one can design a more equitable and ethical storm water pollution control policy than the one currently being followed by the LARWQCB. For instance, the Board could more aggressively pursue the development of a combination of treatment wetlands and infiltration systems. Establishing wetlands and enlarging existing ones in less populated areas of Los Angeles County to treat runoff by natural means would not be very expensive and would create habitats for a variety of plants, birds, and wildlife. A system of wetlands spread across the County would improve the quality of the natural environment and would be enjoyed by residents living near the wetlands.

Creating infiltration systems in the inland-urban areas of the County is another way to disperse equitably the costs and benefits of controlling storm water runoff contamination. In areas of high density, where yards of homes are small and industrial facilities have large roof and parking areas, runoff flow is high. Encouraging homeowners to plant certain kinds of shrubs, bushes, and trees in backyards, possibly through public education and by subsidizing purchasing costs, can have a major combined effect on reducing the amount of contaminated runoff entering storm drains and the ocean. Such an approach would also improve the beauty of neighborhoods and increase property values. Similarly, businesses operating in industrial areas can be required to create sizeable green spaces around their facilities and parking lots to catch much of the storm water flow (as opposed to making more expensive structural modifications). Lawns, bushes, and trees (maintained by natural means) can beautify these areas, improve infiltration, and reduce pollution control costs for businesses.

Furthermore, the County should engage in a massive program to develop large green spaces and parks for surrounding community residents. According to the Trust for the Public Land (2005), Los Angeles County has among the fewest and smallest parks per square mile compared to other counties in the nation. Presently, there are nearly thirty-two acres of park per thousand residents in largely white neighborhoods, compared with less than one acre in Hispanic neighborhoods and just less than two acres in African-American communities (McGreevy 2006). Over 1.5 million children, many living in poor and minority neighborhoods, have no park, playing field, or place to recreate, and about two out of three children in the region do not have a park nearby (Trust for the Public Land 2005). Opening new parks and green playing fields would not only help infiltrate storm water runoff and help replenish underground aquifers, they would also provide large numbers of inner-city youth and adults with a variety of recreational opportunities and pleasant surroundings. The national parks in Southern California are less accessible to inner-city residents and tend to cater to wealthy, white citizens. Although land is expensive, government officials could make a stronger effort to revitalize brownfields and other vacant lands and turn them into public parks and green spaces. Such an approach would increase the price of existing property in these areas and enhance the quality of life of urban-inland families at a relatively low cost.

Conclusion

Future environmental policymakers in Southern California face significant challenges in their attempt to improve water quality and address serious equity and ethical issues. For instance, Southern California continues to experience steady increases in population growth and development, thereby increasing demand for housing. As a result, housing prices have escalated since the early 1990s, and only high-wage earners can now afford to live in most locations. This trend, if it continues, will widen the gap between rich and poor in the region, thereby making it even harder for a number of inland communities, many of which include high percentages of poor and minority residents, to meet the costs of required storm water pollution abatement.

Furthermore, as this chapter suggests, continued population growth and development in the region will add more sources and greater amounts of pollution of various kinds. Existing green spaces that could be used for absorbing storm water runoff will likely be reduced or eliminated as new housing and commercial structures are built. The paving of many miles of asphalt will reduce storm water absorption by the soil. As a consequence, environmental policymakers will be forced to depend more heavily on alternative strategies for improving water quality, which will further drive up the costs of pollution control in the region.

Finally, there is no guarantee that the present approach by the LARWQCB will achieve acceptable water quality standards. Much is still unknown about the level of effectiveness of present abatement strategies as well as the precise health and environmental threats of different storm water pollutants. If the results of ongoing scientific research on the impact of storm water pollutants require policymakers to implement even stricter regulations and standards, the cost of pollution control will dramatically increase. This will present leaders at all levels with greater obstacles to meeting new water quality goals without exacerbating existing equity problems. Whether they will be able to overcome these obstacles and achieve both pollution abatement and social equity is uncertain at this point.

Notes

The authors are greatly indebted to Taylor Dalton for his excellent research assistance.

1. Most of this section is drawn from Devinny, Kamieniecki, and Stenstrom 2004.

2. Although marine scientists cannot be certain, they suspect that the large homeless population, numbering over 90,000, in the region may be the primary source of the human waste.

3. On February 27, 2004, a Los Angeles County Superior Court judge dismissed several claims filed in five lawsuits that would have invalidated regional storm water control regulations as promulgated by the LARWQCB. The lawsuits were filed by the Coalition for Practical Regulation (CPR). See Water Environment

Federation 2004. On March 25, 2005, a Los Angeles County Superior Court judge rejected arguments by Los Angeles County, CPR, and the home-building industry that the LARWQCB ignored economic effects in developing a comprehensive plan to clean up polluted storm water runoff (Bustillo 2005). In June 2005, the Los Angeles County Board of Supervisors voted 3 to 2 to continue the legal battle against the imposition of strict rules on cleaning up storm water runoff (Leonard 2005). (The vote was divided by race, and inland-versus-coastal representation.) The CPR also has sued the EPA over its adoption of trash Total Maximum Daily Loads (TMDLs) in U.S. District Court. In late 2004, voters in the city of Los Angeles overwhelmingly passed Measure O, a bond measure allocating $500 million for storm water pollution control projects.

4. The figures for median house values have risen dramatically over the last five years, particularly in affluent communities such as those that are located near the ocean. Thus, the differences between coastal and other communities are probably even greater now than the table suggests.

5. Lester et al. 2001 believe that social class provides a better explanation for environmental justice than race.

6. For an in-depth discussion of utilitarianism see the works of Bentham (1907), Mill (1979), Moore (1947), and Sidgwick (1981).

7. Interestingly, applying the same logic, the reverse can also be true. One's intentions or the probable outcome of an action could be bad, but because of some unforeseen circumstance, the consequence of his/her action turns out to be good. In this scenario, regardless of its origins, the action would be deemed right. Such logic, however, begs criticism. Should happenstance be able to convert a wrong into a right? Perhaps, then, motive and the anticipated or intended outcome should be considered in determining ethical value.

8. See Adams (1967) for a more thorough discussion of motive utilitarianism.

9. The LARWQCB and the city of Los Angeles have reached a settlement on trash TMDL. Under the agreement, the city will spend $168 million to reduce trash releases by 50 percent over the next five years.

References

Adams, Robert Merrihew. 1967. "Motive Utilitarianism." *Journal of Philosophy* 73:467–481.

Beckman, David S., Mark Gold, and Steve Fleischli. 2002. "Cities Exaggerate Cost of Water Cleanup." *Los Angeles Times,* March 10, http://www.cleanwater-alliance.org/citiesexag.htm.

Bentham, Jeremy. 1907. *An Introduction to the Principles of Morals and Legislation.* Oxford: Clarendon Press.

Bowman, Ann O'M. 1997. "Environmental (In)Equality: Race, Class, and the Distribution of Environmental Bads." In Sheldon Kamieniecki, George A. Gonzalez. and Robert O. Vos, eds., *Flashpoints in Environmental Policymaking:*

Controversies in Achieving Sustainability. Albany, New York: State University of New York Press, pp. 155–175.

Brink, David O. 1986. "Utilitarian Morality and the Personal Point of View." *Journal of Philosophy* 83:417–438.

Bryant, Bunyan, ed. 1995. *Environmental Justice: Issues, Policies, and Solutions.* Washington, DC: Island Press.

Bryant, Bunyan, and Paul Mohai, eds. 1992. *Race and the Incidence of Environmental Hazards: A Time for Discourse.* Boulder, CO: Westview Press.

Bullard, Robert D. 1994a. *Dumping in Dixie: Race, Class, and Environmental Quality.* 2nd ed. Boulder, CO: Westview Press.

Bullard, Robert D., ed. 1994b. *Unequal Protection: Environmental Justice and Communities of Color.* San Francisco: Sierra Club Books.

Bustillo, Miguel. 2005. "Judge Upholds Plan to Clean Up Polluted Runoff." *Los Angeles Times,* March 29, B4.

Camacho, David E. 1998. "The Environmental Justice Movement: A Political Framework." In David E. Camacho, ed., *Environmental Injustices, Political Struggles: Race, Class, and the Environment.* Durham, NC: Duke University Press, pp. 11–30.

Conway, M. Margaret. 2000. *Political Participation in the United States.* Washington, DC: CQ Press.

Devinny, Joseph S., Sheldon Kamieniecki, and Michael K. Stenstrom. 2004. "Alternative Approaches to Storm Water Quality Control." Report prepared for the Los Angeles Regional Water Quality Control Board.

Devinny, Joseph S., Michael K. Stenstrom, and Sheldon Kamieniecki. 2005. "An Estimate of the Total Costs and Benefits of Storm Water Quality Control for the Los Angeles Region." Manuscript.

Dwight, Ryan H., Linda M. Fernandez, Dean B. Baker, Jan C. Semenza, and Betty H. Olson. 2005. "Estimating the Economic Burden from Illnesses Associated with Recreational Coastal Water Pollution: A Case Study in Orange County, California." *Journal of Environmental Management,* http://www.sciencedirect.com/science?_ob=ArticleURL&_udi=B6WJ7–4G0YTMK–1&_us.

Foreman, Christopher H. 1998. *The Promise and Peril of Environmental Justice.* Washington, DC: Brookings Institution Press.

Forester, Larry (Coalition for Practical Regulation Steering Committee). 2004. "Comments on the State Water Resources Control Board's Triennial Review of California Ocean Plan." Letter sent via e-mail to Frank Roddy, Division of Water Quality, State Water Resources Control Board, May 17.

Gordon, Peter, J. Kuprenas, James E. Moore, Harry W. Richardson, and C. Williamson. 2002. "An Economic Impact Evaluation of Proposed Storm Water Treatment for Los Angeles County." Los Angeles: University of Southern California. Report prepared for the Coalition for Practical Regulation.

Kamieniecki, Sheldon, and Janie Steckenrider. 1997. "Two Faces of Equity in Superfund Implementation." In Sheldon Kamieniecki, George A. Gonzalez, and

Robert O. Vos, eds., *Flashpoints in Environmental Policymaking: Controversies in Achieving Sustainability.* Albany, New York: State University of New York Press, pp. 129–154.

Leonard, Jack. 2005. "County to Fight Water Cleanup Rules." *Los Angeles Times,* June 23, B4.

Lester, James P., David W. Allen, and Kelly M. Hill. 2001. *Environmental Injustice in the United States: Myths and Realities.* Boulder, CO: Westview Press.

Los Angeles Times. 2005. "AutoZone Accused of Dumping in California," June 23, B6.

McGreevy, Patrick. 2006. "Audit Details Neglect of Parks." *Los Angeles Times,* January 10, B3.

Mill, John Stuart. 1979. *Utilitarianism.* Indianapolis: Hackett Publishing Company.

Mohai, Paul, and Bunyan Bryant. 1992. "Environmental Racism: Reviewing the Evidence." In Bunyan Bryant and Paul Mohai, eds., *Race and the Incidence of Environmental Hazards.* Boulder, CO: Westview Press, pp. 163–176.

Moore, George Edward. 1947. *Ethics.* London: Oxford University Press.

Morgenthau, Hans J. 1945. "The Evil of Politics and the Ethics of Evil." *Ethics* 56:1–18.

O'Leary, Rosemary. 2003. "Environmental Policy in the Courts." In Norman J. Vig and Michael E. Kraft, eds., *Environmental Policy: New Directions for the Twenty-First Century.* Washington, DC: Congressional Quarterly Press, pp. 151–174.

Pendleton, Linwood. 2001. "Managing Beach Amenities to Reduce Exposure to Coastal Hazards: Storm Water Pollution." *Coastal Management* 29:239–252.

Pendleton, Linwood, Nicole Martin, and D. G. Webster. 2001. "Public Perceptions of Environmental Quality: A Survey Study of Beach Use and Perceptions in Los Angeles County." *Marine Pollution Bulletin* 42:1155–1160.

Public Policy Institute of California. 2003. "It's a Beach State . . . of Mind: Despite Tumultuous Times, California's Golden Coast Still Captures Hearts." Press release, http://www.ppic.org/main/pressrelease.asp?i=461.

Riley, Jonathan. 1990. "Utilitarian Ethics and Democratic Government." *Ethics* 100:335–348.

Ringquist, Evan J. 2003. "Environmental Justice: Normative Concerns, Empirical Evidence, and Government Action." In Norman J. Vig and Michael E. Kraft, eds., *Environmental Policy: New Directions for the Twenty-First Century,* Washington, DC: Congressional Quarterly Press, pp. 239–263.

Schlosberg, David. 1999. *Environmental Justice and the New Pluralism.* New York: Oxford University Press.

Sidgwick, Henry. 1981. *The Methods of Ethics.* Indianapolis: Hackett Publishing Company.

Simpson, Andrea. 2002. "Who Hears Their Cry? African-American Women and the Fight for Environmental Justice in Memphis, Tennessee." In J. Adamson, M. M. Evans, and R. Stein, eds., *The Environmental Justice Reader: Politics, Poetics, and Pedagogy.* Tucson: University of Arizona Press.

Singer, Marcus G. 1977. "Actual Consequence Utilitarianism." *Mind,* new series, 86:67–77.

Sutherland, Ronald, J., and Richard G. Walsh. 1985. "Effect of Distance on the Preservation Value of Water Quality." *Land Economics* 61:281–291.

Trust for the Public Land. 2005. http://www.tpl.org/tier3_cdl.cfm?content_item _id=15115 &folder_id=2627.

Verba, Sidney, and Norman H. Nie. 1972. *Participation in America: Political Democracy and Social Equality.* New York: Harper and Row.

Water Environment Federation. 2004. "Court Dismisses Claims Filed by Opponents of Southern California Storm Water Rule." http://wef.org/MemberZone/ WefReporter/2004/wefreporter.0310.jhtml.

4

Equity and Water in Mexico's Changing Institutional Landscape

Margaret Wilder

As nations across Latin America and the developing world turn to market-oriented restructuring of their water policies, the question arises: what are the implications for equity in the distribution, use, and access to water for marginalized sectors? Water is increasingly recognized as a scarce resource that is critical to global poverty reduction and health improvement strategies (United Nations 2005). Developing nations, under the tutelage and funding of international institutions such as the World Bank and the Inter-American Development Bank, have restructured the "architecture of governance" in recent years in ways that have major implications for equity in the use and distribution of water resources (Ingram et al. 2001, 9). Reform packages are being superimposed on Latin American landscapes that are pockmarked by the highest levels of inequality of any world region and which seemed impervious even to declines in poverty levels.[1]

Countries in Latin America including Brazil, Colombia, Chile, Perú, Nicaragua, Costa Rica, and Honduras have all instituted water management reforms with some similar elements (Lemos, this volume; Bauer 1997; Easter et al. 1998). In this chapter, I examine the new institutional landscape for managing water in both urban and rural Mexico in terms of the equity implications of these changes. Mexico's water sector "modernization" was embedded within a widespread program of neoliberal economic restructuring during the 1990s that promoted market mechanisms, trade liberalization, and state retrenchment. This restructuring affected every economic sector. Mexico's reform package had two key philosophical thrusts associated with its broad program of decentralized management of productive resources and privatization of resources, involving enhancement of what I will term "political equity" on one hand and "economic equity" on the other. In this chapter, political equity refers to the institutionalization of local participation in water policy-making as well as the quality of participatory mechanisms; on the other hand, economic equity refers to a range of socioeconomic aspects of water, including affordability and productivity, as well as accessibility.

In this chapter I examine the changes in equity in three distinct areas of institutional transformation based on water-sector research in the northern Mexican state of Sonora: urban water management, river basin councils, and irrigation districts. Why is the Sonoran experience a key indicator of the implications of the emerging water management paradigm in Mexico? Sonora is the most highly irrigated state in Mexico, with seven large irrigation districts producing commercial export crops, including fresh table grapes, citrus, and asparagus, as well as wheat and raisins for the domestic market. Sonora is also part of the most drought-prone (northwest) region of the country (CNA 2001), and has been in a severe drought since the mid-1990s. Finally, demographic pressures resulting from Sonora's rapidly growing population are anticipated to create additional stress on water supply in both the near term and the long term, particularly in the context of climate variability, periodic drought, and expected climate-change impacts. Sonora is arguably one of the Mexican states that is best integrated into the global economy, based on its industrial base, commercial agricultural production, transportation network, and the cultural and economic interdependence of the border region. Given these characteristics, Sonora, like other industrialized areas of the north and center regions of Mexico, was well positioned to benefit from the neoliberal economic restructuring and modernization of the water sector, as represented by the new governance regime. Sonora is therefore a critical laboratory in which to examine the equity implications of transitions in water governance.

Political aperture, a professed interest in public accountability and governmental transparency, and the requirements of the new water management paradigm promoted by the World Bank, were all factors that led to formalizing public participation mechanisms within the emerging water management institutions. Via the National Water Law, Mexico's water policy was reshaped from a clientelistic, state-centered framework to a decentralized, user-centered model of management, with authority over irrigation districts transferred from the federal government to local water user associations. In urban areas, water service management was devolved from federal to state and municipal levels by the water reforms (Wilder and Romero Lankao 2006).

The water reforms created a host of new management institutions to take on the task of decentralized management, including river basin councils to practice integrated water resource management and engage in sustainable, long-term planning at the basin level; water user associations (managed by professional, highly trained engineers) to manage water supply and delivery within major irrigation districts; and "civil associations" comprised of irrigator groups to manage water at the subdistrict level. I use economic equity to refer to a range of factors including availability, accessibility, affordability and, in the case of agriculture, productivity of water—who benefits from the productivity generated by water? National economic restructuring changed the nature of water economics through institution of such principles as full-cost recovery water pricing and restructuring of water governance such that the state has retrenched its water subsidies and plays a circumscribed financial role.

In this chapter, I argue that these two strains of political and economic equity have coexisted in generally contradictory ways over the last dozen years. Emerging forms of water governance represented by decentralized water management embedded within the nexus of neoliberal economic reforms have resulted in a discordant and inharmonious melody. Such a melody only minimally realizes the promises associated with decentralization while at the same time driving downward economic outcomes, especially for marginalized populations. The first section of this chapter details Mexico's water reform program and the uneven landscape of water access and management at the outset of the reforms, in the early 1990s. The second section analyzes the concept of political and economic equity, drawing on recent theories relating to water and equity. In the third section, I present an overview of findings from three in-depth studies from urban areas, irrigation districts, and river basin councils in Sonora. In the last section, I discuss the equity implications of these findings for marginalized sectors of Mexican society.

I

In 1989, the administration of President Carlos Salinas de Gortari of the long-ruling Party of the Institutionalized Revolution (PRI) created a new federal government agency to implement a program of water reforms that would reflect Mexico's "new water culture" (*la nueva cultura de agua*). Salinas' reform of the water sector mirrored the transformation of Mexico's economic model to a neoliberal, market-oriented framework based on opening up markets, eliminating trade barriers, and entry into the World Trade Organization. Mexico's water reform program was shaped and partially financed by the World Bank, which conditioned its financial aid on establishing decentralized water management mechanisms and introducing privatization components into the operation and management framework. In 1992, the new water agency, the National Water Commission (CNA, Comisión Nacional de Agua) developed a National Water Law (Ley de Aguas Nacionales) that instituted a sweeping set of reforms and created new subnational-level mechanisms for managing and operating major water systems, including urban water services and irrigation water. Through the new water legislation and its attendant regulations, Mexico's water policy was reshaped from a state-centered to a user-centered model of management, with authority over irrigation districts transferred to local water user associations from the federal government. In urban areas, water service management was, in theory, decentralized from federal to state and municipal levels by the water reforms (Wilder 2005).

Urban water systems in Mexico are "in transition" between a subsidized service provided by the government and a self-sufficient autonomous model. Under decentralization, urban water managers have largely inherited deteriorated systems close to the end of their useful lives, marked by a lack of water metering and a high degree of water loss in the systems, and are scarcely able to cover operational costs with the water fees collected. The water reforms created a host of new management institutions to take on the task of decentralized management, including a national network of twenty-five river basin councils to influence sustainable, long-term planning at the basin level; water user associations to manage water supply and delivery within eighty-one irrigation districts; and "civil associations" (also known as water modules) to make decisions about water at the sub-irrigation district level. Like the urban water systems, the irrigation districts prior to transference were also seriously deteriorated with systemic losses of nearly 50 percent and highly subsidized when transferred to water users in the mid-1990s (Wilder 2005). In April 2004, under the Vicente Fox administration, the national water law was updated to include an enhanced focus on decentralized governance, local participation, integrated watershed management, and environmental sustainability, but the regulations to implement the changes have not yet been adopted.

Mexico's water reform program was superimposed onto a landscape of highly uneven distribution of water stemming from both natural systems and political structures and inequitable distribution of access to water for domestic purposes and sustenance of livelihoods. Owing perhaps to a kind of cosmic perversity in combination with poor planning and erratic lifestyle choices—similar, indeed, to irrational growth patterns we witness in the sunbelt region of the arid southwest U.S.—Mexico has water in abundance in less productive, slow-growth regions of the south and southeast where it is least needed to meet domestic and economic needs, while the burgeoning center and north, host to industrial growth and vast rural-urban migration streams, cannot meet the rampant demand.[2]

The highest growth areas and most productive regions of Mexico's center, north and northwest have dangerously low per capita levels of water supply (CNA 2001, 23). The north and northwest regions of Mexico have been subject to a prolonged drought over most of the past decade, resulting in federal declarations of "states of emergency" at various times during the period (Becerril, Muñoz, and Camacho 1999). Agriculture uses about 80 percent of total water supply in Mexico, and irrigated commercial agriculture is concentrated in the center and arid northwest states. Sonora is the most highly irrigated state in the country, with nearly 10 percent of Mexico's irrigation districts (CNA 2001). Declining water quality, due to lack of municipal and industrial treatment, and growing salinization of coastal aquifers are growing problems. The dearth of municipal water treatment facilities has led to health problems resulting in illness and high child mortality rates.[3] Residential water and sewerage service is highly uneven between urban and rural areas, as well as between better-off and poorer colonias within urban areas.[4]

Irrigation is another area where inequities exist in the waterscape. Major irrigation districts are concentrated in the northwestern region and central regions (Sonora, Sinaloa, Baja California Norte); by contrast, the southern states (Chiapas, Quintana Roo, Yucatán, Oaxaca) have the lowest percentages of irrigated area. The land tenure pattern across Mexico also adds to the landscape of inequity, as most of the productive land is in large, private parcels, with less productive and smaller marginal lands allocated to 3 million ejidatarios. Land tenure liberalization instituted in 1992—known as the ejido reforms—allowed communal farms, or ejidos, for the first time legally to rent or to sell their land following a process of titling and certification (Cornelius and Myhre 1998; Wilder

2002; Wilder and Whiteford 2006). Regional imbalances in the natural supply of water and the human demand on existing water supply, coupled with regional climatic variations[5] and socioeconomic and structural factors driving growth, results in a pattern that caused Búrquez and Martinez Yrizar to comment that Mexico's natural distribution of water supply results in "a profound division between a 'south' rich in water and a 'north' subject to chronic scarcity" (Búrquez and Martinez Yrizar 2000, 277). In Mexico, then, natural climatic and geologic conditions combine with human institutions and structural processes to form an uneven landscape of access to water among and within specific regions, a situation that has exacerbated and accelerated the impacts of the sweeping water reforms. The next section considers theoretical frameworks in which to evaluate linkages between water and equity.

II

Exploring the concept of equity within the context of water management is an interpretive and subjective exercise, yet, an increasingly significant one given the growing competition over scarce water resources and the importance of water in contemporary socioeconomic activities (Prasad et al. 2006, 63). Equity in water management and allocation is notoriously difficult both to define and to measure (Tisdell 2003; Prasad et al. 2006). The role of the state and the role of water users in managing water also have equity implications, thus bringing into relief not only distributive water allocation formulae but also legal frameworks and mechanisms for making policy and allocation decisions that are responsive to changes in water supply (for example, in cases of drought and water scarcity) (Meynen and Doornbos 2004).

The demands of environmental sustainability (if understood to have a future-oriented component, as per the Brundtland definition[6]) and those of social equity (that can have either a contemporary or a future orientation) are not always consistent and can therefore result in contradictory outcomes. (Wilder and Romero Lankao 2006; Meynen and Doornbos 2004). As Prasad and his coauthors argue, the concept of equity is "diverse, complex, and contextual" (Prasad et al. 2006, 65). They outline four distinct perspectives that can be used to analyze equity in a water resource context: access (both physical and economic) to the resource; distribution of socioeconomic impacts; extent of use; and cost-sharing (Prasad et al. 2006, 65). Tisdell examines water equity from the point of view of social justice theory, drawing on Rawls, Bentham (utilitarianism), and Nozick to critique three principal water doctrines in use today: the riparian doctrine, the prior appropriation doctrine, and the "nonpriority permit" doctrine (Tisdell 2003, 405). Meynen and Doornbos (2004) discuss water equity and sustainability in terms of their management and policymaking implications, considering these questions in light of decentralized water management institutions. They argue that institutional arrangements for making water policy are key to assessing the underlying (or overarching) equity considerations. In short, the role and relative quality of "community participation" in decision making about water use and allocation is directly constitutive of whether or not water policy can be considered equitable. As the authors point out, the quality of public participation and the avoidance of elite capture of participatory processes is an important caveat to an otherwise broadly sound principle.

Drawing on this finely textured and richly complex theoretical backdrop, in this chapter I intend to sidestep finer distinctions and to discuss just two broadly drawn categories—political equity and economic equity. I argue that these categories bring together the traditional distributional concerns of water and equity discussions (à la Prasad et al. 2006 and Tisdell 2003) with a focus on procedural characteristics based on the degree and quality of local participation. By political equity, I refer to the role of local participation in water policy construction, regulation, and implementation. Political equity is a composite term intended to capture the democratization of water policy development via the formalized participation of local water users, user management, enhanced local control and greater reliance on local knowledge. On the other hand, economic equity refers to the availability, accessibility, affordability, and productivity of water, including the question of who benefits from water's generative activities.

As Prasad et al. (2006) make clear, water must remain affordable to qualify as accessible. Similarly, water must be safe and of potable standards in order to satisfy equity measures. Urban and rural residents should have residential water and sewerage service that is consistently available at affordable rates. In farming areas reliant on irrigation water, water must be available and affordable to all irrigators within the district, regardless of land tenure status (e.g., private producer or *ejidatario*), and water allocation schemes among land tenure sectors should be standardized and nonpolitical. Equity is not always a function only of water policies themselves—since water is embedded within a sociopolitical framework in an increasingly globalized economy, multiple stresses and factors such as trade liberalization and water markets also create impacts within the water sector and among water users.

Based on the Mexican case, I argue that different forms of equity coexist within Mexico's "new culture of water." It is essential to differentiate between different shades of equity, since political equity has been enhanced by Mexico's dramatic water reform program while economic equity has remained the same or declined under the water reform package, resulting in a situation where opportunities and platforms for engaging in water policy-making are emerging even as the associated economic opportunities are closed down. Some scholars have argued that in recent decades there has been close to a wholesale shift from a global water policy based on social equity to one based on economic efficiency (see discussions in Bakker 2005; Gleick et al. 2002).

This chapter advances a more nuanced and hybrid view that political and economic equity coexist in the Mexican case in often uneasy and contradictory ways: gains in political equity represented by codified, formal opportunities to participate in water policy decisions have been more than offset by adverse changes in economic equity related to liberalized markets, loss of subsidies and consumer-pays water pricing principles. On balance, the net change for water users *has* been a negative one, resulting in a net downward shift in economic equity, even as political equity has formally expanded to a small extent, creating important democratic spaces that have potentially significant implications for equity and sustainability. In practice, the political equity gains have not (or not yet) been realized due to lack of meaningful implementation and circumscribed formal authority of emerging water governance institutions. The dynamic northwestern state of Sonora represents the kind of region that might have been expected to benefit from Mexico's modernization and water reform strategies, yet the evidence from Sonora indicates a mixed record in equity terms. The next section examines the question of water and equity in Sonora in three particular contexts: urban, irrigation, and river basin councils.

III

Sonora is one of the most significant states in Mexico's northwest due to its location on the border with the United States, its role as the principal transportation corridor for export agriculture, particularly fruits and vegetables from Sinaloa and Sonora to the western United States, and its dynamism in terms of industrial and population growth. The Sonora-Arizona region is one of the principal crossing points for both legal and undocumented migrants along the Mexico-U.S. border, and Mexico's industrialization strategy focused traditionally on foreign-owned export assembly plants (*maquiladoras*) in the border zone has led to faster-than-average population growth and relatively dense concentrations of maquiladora plants. These processes have significant implications for water use, supply and demand, and quality. As the most highly irrigated state in Mexico, with nearly 10 percent of all irrigation districts, Sonora also demonstrates the extreme unevenness of development in an acute way. The border between the two countries is unique in linking two countries with vastly disparate economies; but Sonora itself also exhibits an extreme bifurcation between the highly developed, technified, and urbanized western coast and the traditional, rural *serrana* region running lengthwise along the Sierra Madre range in eastern Sonora (West 1993). Given its high degree of dynamism, its significance as an economic and geopolitical space, its natural desert aridity, and the role it plays in terms of migration streams, urban growth, and both agricultural and industrial development, Sonora represents a key laboratory in which to examine processes relating to water, equity, and uneven development. Next, I discuss three case studies—urban water management, river basin councils, and irrigation districts—to analyze the salience of water and equity issues under these lenses.[7]

Urban Water Management

Urban water service delivery in Mexico has been decentralized[8] under the National Water Law from being a federal government responsibility to the state and municipal levels. The decentralization impulse stems from transformations in philosophy in international institutions such as the World Bank, which supports decentralization as an essential aspect of water sector modernization, especially for developing countries, and conditions its financing packages on decentralizing reforms. Larson and Ribot (2004) have argued that there is a changing "language of decentralization" with respect to environmental resource management, evidenced by a transition from "efficient management" to an entirely new lexicon involving "a new emancipatory language of democracy, pluralism, and rights."

The promises associated with decentralization include greater public participation, utilization of local knowledge, improved efficiency, poverty reduction, and more sustainable resource management (Larson and Ribot 2004). The benefits of decentralization and local participation are reified in international water doctrines and policies (e.g.,the Dublin Principles, World Bank water policy, UN Millennium Development Goals). These principles implicitly claim that decentralized water management will increase equity by conferring economic benefits (such as poverty reduction) and by involving local communities in decision-making processes. Sonora's experience with urban water management decentralization can yield a number of conclusions relating to equity.

In a study of ten major urban areas in Sonora, we found substantial heterogeneity of institutional arrangements, including municipal, state, private concessions, as well as evidence of transboundary (U.S./Mexico) cooperation (Wilder in submission). Although community participation mechanisms are formalized within the legal framework, we found scarce evidence that either formal—or even informal—mechanisms exist and are integrated into decision-making processes in practice. In limited cases, public participation has been formalized, such as in the state capital, Hermosillo, where the municipality formed a technical/citizens' advisory council on water; but outside Hermosillo, formal processes to involve the community in water policy making were, until recently, either weak or nonexistent (Wilder in submission).

Since the passage of a 2006 state water law that required formalized local participation in municipal water policy, many Sonoran municipalities have followed Hermosillo's lead and created an advisory council. However, these meet infrequently and have not yet crystallized into effective bodies to substantially influence policy. One important exception to this conclusion—albeit not one associated with decentralization, per se—is the very successful public participation components in border communities working with the Border Environment Cooperation Commission (BECC). BECC was created by the NAFTA side agreement on environment to certify water and wastewater projects in border communities. BECC has an open, transparent and participatory set of processes that has been heralded as one of the principal achievements of the organization, and which may be credited with helping to instill a "new culture of participation" in decision making about water, at least within border communities (Lemos and Luna 1999; Varady et al. 1997).

Pineda (2006), Wilder and Romero Lankao (2006) and others have found that municipal and state water managers, in Sonora and other regions of Mexico, lack the necessary financial resources, compared with the federal government, to manage and operate complex urban water delivery systems that rely on outdated infrastructure requiring investment and repair. Despite efforts to create a new culture of water, there has not yet been created a culture of willingness for water customers to pay higher water tariffs, nor for efficiencies in water accounting. Water tariffs are set to cover the accumulated past debt, not to meet future needs, and fee rates are based on what can feasibly be asked rather than established to cover actual costs (Pineda 2006, 15).

Some customers pay; others continue as freeriders. But, overall, the current tariff system is inequitable(Pineda 2006). Inequities in service provision across space and socioeconomic context persist, although they are less pronounced in Sonora than in more rural and isolated parts of Mexico, such that many poor colonias do not have residential water service. Even the majority of neighborhoods that do have residential water service actually have access to tap water only hours per day, due to drought and water shortages. The decentralized management of urban water services has opened new vistas of possibility and promise for gains in political equity, but only in limited cases, such as that of Hermosillo, have there been verifiable gains in formalized community participation. The successful participation strategies practiced by BECC in working with border communities may be a model that Sonoran cities could follow to enhance political equity.

Incomplete provision of residential water service, lack of access to water 24 hours a day, and poor quality drinking water all lead to declines in economic equity, in that households spend larger proportions of income on buying bottled and trucked-in water than if residential water service were expanded, accessible, clean, and reliable. These problems did not develop under decentralized management structures for water, but they also have not yet been effectively addressed by the new institutional arrangements. In fact, there is good evidence that municipalization is often a recipe for privatization—in the form of a private company concessionaire —leading to higher water tariffs. Municipal governments with very limited technical and financial resources often have little recourse but to turn to a private concessionaire for relief (Pineda 1999).

River Basin Councils
Mexico has a national network of twenty-five major river basin councils, complemented by networks of commissions and committees at the subwatershed level.[9] The river basin councils have voting membership as

follows: the director of the National Water Commission (designated as the chair of the river basin council); the governor of the state(s) in which the watershed lies; and an elected representative from each of the uses existing in the watershed in number at least equal to the government representatives, including agriculture, agro-industry, domestic, aquaculture, services, industrial, fisheries, and urban public. In addition, there is theoretically a representative allowed "for ecological conservation" (SEMARNAP 1998, 31). In addition, other representatives are allowed to participate in discussions (e.g., *voz*) but are not allowed a vote (*voto*): other federal and state government agencies, municipal councils, nongovernmental organizations, and academics. The technical ministry of the National Water Commission also has a voice-only participation (SEMARNAP 1998). Although the river basin council is intended to elicit citizen input and user participation, its formal membership weighs heavily toward government representation, including the president of the river basin council who also has the power to set the agenda and represent the group publicly.

In Sonora, three river basin councils on the state's three major watersheds were officially established in 1999 and 2000, but did not have active memberships and meetings until approximately 2004.[10] Major revisions in April 2004, to the National Water Law created a new "regionalized" structure for the National Water Commission, and strengthened the focus on water user participation and the river basin councils. In theory, the river basin councils represent a progressive step forward in using integrated water resources management strategies to find workable solutions, conduct long-term planning, and develop environmentally sustainable outcomes within a watershed. In reality, we found logistical and jurisdictional obstacles to effective functioning of the river basin councils, and implementing regulations for the 2004 changes have not yet been adopted.

The composition of the major river basin councils is so comprehensive that the river basin council coordinators complain it is difficult even to schedule a meeting of the full group. In particular, it is very challenging and probably impractical to find meeting times for the governor himself to attend. More significant is the councils' lack of formal jurisdiction under the legal framework setting forth their charges and responsibilities. For example, river basin councils do not have formal authority to redistribute water allocations among different user groups, limiting the impact they can have on any given problem. The lack of jurisdiction, in turn, makes council participants less willing to give time to a process that has only an insubstantial and advisory nature. How "representation" is constructed for the river basin councils is another problem. Local irrigators, in particular, are concerned that agriculture, which uses 80 percent of available water, holds only one vote on the river basin councils, as do much smaller water-user sectors. Government representation is disproportionately high, compared with very limited participation from the citizenry. Procedural transparency and accountability were largely lacking, and there is a lack of openness about meeting agendas, meeting minutes, and intolerance of meeting attendance by noncouncil members. Marginal groups are not represented on river basin councils. The urban poor, the *colonia* residents without residential water or sewerage service, and the *ejidatario* (small-scale communal farmer) do not have a seat at the council table, pointing to major shortcomings in terms of economic equity. By definition, the process only allows representation for water-using sectors; thus, those who are still disenfranchised from the water system will not find any means of redress within the river basin council framework.

These findings suggest that river basin councils and their related council networks at the subwatershed level have unrealized potential for serving as an important site of collaborative citizen and state engagement with resolving water problems and planning for long-term sustainability at the watershed level. The creation of river basin councils led to formal gains in political equity, by codifying citizen participation into new processes and structures under the water reform program, but those changes toward political equity have not yet been made meaningful. Meynen and Doornbos (2004), Larson and Ribot (2004), and others have argued that local participation in environmental resource management must be more than just the formal trappings of participation in order to be effective and representative. Economic equity would dictate that Mexico's participatory strategies be modified to include a formalized role for marginalized groups who are not represented by the current structures and a meaningful, influential role and agenda in water policy making.

Irrigation Districts

A 1998 World Bank report heralds Mexico's irrigation transfer program as "the most successful of the new globalizers" (Easter et al. 1998). Of what might be termed the three principal "experiments" in decentralized water management reform addressed in the National Water Law and discussed in this chapter—at the urban, watershed, and irrigation district levels—the transference of irrigation districts is arguably the only one that, to date, has truly increased political equity. Prior to transference, the federal government agriculture and water agencies would come into districts and dictate planting regimes, amounts of available irrigation water, and establish irrigation fees to be paid. Such fees would be sent back to Mexico City and held in general revenue pools, almost certainly not to be used for improvements within the districts where they were generated.

The 1992 law transferred responsibility for managing Mexico's major irrigation districts to the water users themselves. Nearly all of the major irrigation districts, including six of the seven districts within Sonora, have been successfully transferred.[11] Irrigation districts have their own elected officers and paid professional staffs at the district level. Each district is divided into water modules (also known as civil associations) which have their own elected representatives at the subdistrict level.[12] The presidency of each water module alternates annually among the private producer, *ejidatario* (small-scale communal producer) and *colono* (small-scale private producer) groups to ensure that more marginal groups are represented within module-level decisions, such as where to make infrastructure improvements and what water fees to establish. Water module civil associations meet frequently, and the irrigation district holds district-wide meetings for all water users at least once a year (Wilder 2002). Producers in the transferred irrigation districts of Rio Yaqui in southern Sonora and Altar-Pitiquito-Caborca in northwestern Sonora testified to high satisfaction with the transference, their level of policy engagement, the functioning of the irrigation district management, and the representativeness of structures in place (Wilder 2000; 2002).

Beyond the gains in political equity represented by the transference, however, the outcomes of the nexus of economic liberalization strategies enacted by Mexico in the 1990s suggest a much gloomier reality. Privatization provisions included in both the water and ejido reforms were, like the decentralization and land-titling initiatives themselves, promoted and financed by the World Bank, as a way for ejidatario producers to gain secure recognition of their land and water rights, in order to then participate in land and water markets via sales or legal rentals of their productive resources. While many, perhaps the majority, of observers viewed these reforms as a thinly veiled path to fast-paced privatization of ejido resources, the intellectual authors of the reform package argued that the economics of ejido production was broken and had to be addressed through modernized market mechanisms that would allow the most entrepreneurial of ejidatarios to take advantage of new market opportunities being opened through Mexico's numerous free-trade agreements. Thobani (1997), a World Bank director of Latin American poverty reduction programs, argued that tradable surplus water rights are a tool in the producer's toolbox to supplement his income when he might otherwise be unable to put his surplus rights to a productive use.

The agricultural modernization program embodied in the three areas of policy and legislative reform—free-trade agreements, ejido reforms, and water reforms—were promoted as a means for the capable and entrepreneurial producer to succeed by participating fully in the competitive play of the free market, with all the tools represented by secure land assets and recorded water rights. The reforms, in other words, were promoted as a way to enhance equity in rural Mexico's productive landscape, by providing to peasant producers the tools of engagement in the capitalist marketplace.

The North American Free Trade Agreement (NAFTA) has had well-documented impacts on small, subsistence producers in southern and central Mexico by opening the floodgates for imported (and subsidized) wheat and grains from Canada and the United States. Less recognized is the impact of 11 other free trade agreements that Mexico entered into over the last dozen years, including those with the European Union, Japan, Chile, Israel and Central America, among others.

The onslaught of free-trade agreements combined with lack of rural credit, state technical assistance retrenchment, and loss of financial subsidies has had a devastating effect on ejidatario producers who have always operated at the margin. A prolonged drought and intensive water use have combined to create water shortages causing irrigation managers to force land out of production, shorten growing seasons to one a year, and retire groundwater wells (Wilder and Whiteford 2006). The net result has been a large-scale abandonment of active production in favor of ejidatarios' renting or selling their land and water rights to better-off private producers or corporations (Wilder and Romero Lankao 2006; Wilder and Whiteford 2006). In two Sonoran irrigation districts studied, one-half to three-quarters of the ejido lands were rented out to private or corporate producers, and ejidos had accelerated marketing of their water rights through a variety of long-term and temporary mechanisms such as annualized water transfers and long-term well rentals that brought remuneration sufficient only to make a payment on accumulated debt at the rural credit bank (Wilder and Whiteford 2006).

In the case of Caborca, intense economic pressure for increasingly scarce groundwater pumping rights was generated by the deep-pocketed asparagus industry, dominated by a few huge corporate growers selling under exclusive contract to transnational companies like Dole, Inc. and Lee Brand (Wilder and Whiteford 2006). Under the weight of such pressures, a kind of "hidden privatization" of communally held land and water rights has occurred that seems to remain under the radar of World Bank researchers and others who track only formal land sales in formal markets, but neglect to see the complex array of informal and (long-term) temporary arrangements that effectively privatize ejidatarios' natural resources (land and water) (Wilder 2002; Wilder in submission).

In the irrigation districts, then, Mexico can point to real gains in political equity in an overall process of management by water users that, at least in the Sonoran cases studied, is highly regarded by water users themselves. These positive effects in terms of political equity, however, are more than offset by the drastic decline in economic equity as the ejido sector struggles to remain in active production under a nexus of economic pressures including trade liberalization, lack of subsidies and government bank credit, and a growing indebtedness that, for many, is insurmountable. Although they have the requisite tools to succeed— including irrigation, technological sophistication and knowledge, integration into commercial export economies, and a location just miles away from Mexico's largest trading partner, the U.S.—these small ejido farmers are unable to compete with well-subsidized U.S. and Canadian farmers.

Implications and Conclusions

Equity is a complex and multifaceted concept that resists easy definition. In the context of Mexico's dramatic program of water reforms initiated in the early 1990s and still under construction today, the concept of equity must be prised into its component parts in order to see its relevance in particular contexts. In this chapter I have introduced the concepts of political equity and economic equity to suggest that these two strains of equity have coexisted in generally contradictory ways over the last dozen years. For example, within the rural context, the water reforms resulted in new forms of political equity represented by greater democratization and participation on the part of local producers and water users who in the past had traditionally been excluded from CNA's tightly controlled federal water policymaking processes.

The water reform package and ejido land reforms ostensibly created more protection for ejidatarios with rights to land and water through a program of land titling and certification and through creating formalized water markets and water banks. Yet under the pressures of the nexus of water and land reforms, productive ejido assets of water and land have increasingly moved into private control (Wilder 2000; Wilder and White-ford 2006). While there have been gains in political equity, there have been losses in economic equity, and the net result has been a separation of ejidatarios from their productive resources.

One might critique the irrigation district analysis by arguing that the negative impacts are created through a nexus of wide-ranging reforms and economic restructuring, and are not solely attributable to the water reforms themselves. However, water resources management and water use are always embedded within a complex network of social, political, and economic relations. In the Mexican case, the water reform program was part of a particular modernization imagery that fundamentally re-structured all significant economic sectors to bring them into alignment with explicitly neoliberal principles. By following what Swyngedouw (2004) terms the "hydrosocial flows"—that is, the ways in which water flows both literally and figuratively map out flows of social power—it is possible to tease apart how the interplay of factors, including the water reforms, shape local impacts and outcomes.

The emerging river basin councils represent a potential gain in po-litical equity for formerly marginalized water-using sectors, although the promise they embody has yet to be realized. The lack of representation of marginalized water users (such as poor colonia residents) is an indication that river basin councils are unlikely to lead to upward shifts in economic equity. In the case of urban water management, decentralized water man-agement has led to some small gains in political equity—for example, in the Sonoran capital Hermosillo where the mayor has formed a citizen/ technical advisory council to advise him on water policy—but the eco-nomic challenges of municipal water management are vast given the re-trenchment of the federal government and the enormous financial burden municipal water service represents. Overall, the discourses of decentrali-zation and the narratives of local empowerment embodied by Mexico's reform package have contradictory outcomes, delivering political power to local water users and land managers with one hand while wringing the economic lifeblood out of the marginalized sectors with the other.

Mexico's political transformation and the movement toward consoli-dation of democracy represented by opposition party gains in the late

1980s and throughout the 1990s, and fulfilled in the 2000 election of opposition (PAN) candidate Vicente Fox, represent perhaps the most transformátive potential for reshaping not only water but environment policy more broadly. The aperture in political control mechanisms formerly tightly held by the PRI have led to unprecedented, yet modest, increases in the role of nongovernmental organizations, growth of environmental activist groups, and the expectation of greater transparency and accountability on the part of the Mexican public.

The retrenchment of the state under the market-oriented economic reforms of the 1990s has created political space for new actors in the water/environmental policy arena. The 2006 election of Felipe Calderón continued the administration of the conservative National Action Party (PAN) and may, in time, make clear whether the new institutional arrangements for water management represented by the cases I have examined here—decentralized urban water service provision, new river basin councils, and user-led irrigation districts—have become a permanent part of Mexico's resource management landscape or were merely an evanescent—or cynical—experiment.

Notes

1. According to ECLAC (Economic Commission on Latin America and the Caribbean), poverty decreased by 3.3 percent from 2005 to 36.5 percent in 2007, (ECLAC 2007). The relationship between poverty and inequality is complex, but two studies found that higher rates of economic liberalization are associated with growing inequality (and, some claim, poverty) (see Huber and Solt 2004; Walton 2004).

2. Mexico's northwest, north, and central regions have 32 percent of water supply coupled with 77 percent of population, and generate 86 percent of GDP, while the south and southeast regions have 68 percent of water supply, coupled with only 23 percent of the population and 14 percent of GDP (CNA 2001, 24; Whiteford and Melville 2002, 3).

3. Nearly one-quarter of total water sources are graded by the CNA as contaminated or highly contaminated, and another 49 percent graded as slightly contaminated; only 22 percent ranked as of "acceptable" quality. Water quality has regional implications as well. The number of severely overdrafted groundwater aquifers with major salinization problems more than tripled from 1975 to 2000 (from 32 to 96 aquifers affected), with the worst cases located in the northwest coastal regions of Sonora and Baja California Sur and Norte, as well as along the Veracruz and Colima coasts (CNA 2001, 28). Whiteford and Melville cite multiple health problems associated with waterborne diseases and which primarily

affect poor Mexican citizens without access to safe drinking water, and resulting in high rates of child mortality and illness (Whiteford and Melville 2002, 6). Mexico had only 793 municipal wastewater treatment systems in operation in December 2000 (CNA 2001, 35).

4. In 2001, the CNA reported rural water service coverage at only 68 percent nationally and sewerage at 37 percent, compared with 88 and 76 percent respectively for urban areas (CNA 2001, 33).

5. Certain climatic patterns such as variable interannual rainfall are associated with the "North American Monsoon region" of northwest Mexico and the southwestern-western United States.

6. According to the 1987 Brundtland Commission (also known as the U.N. World Commission on Environment and Development) report entitled *Our Common Future,* the term "sustainable development" means ". . . to meet the needs of the present without compromising the ability of future generations to meet their own needs."

7. The decentralization and urban water management study was funded by the National Oceanic and Atmospheric Administration (NOAA) under the auspices of the Climate Assessment of the Southwest program, a Regional Integrated Science Assessment Program at the University of Arizona. Between 2003–2005, and with selected follow-up interviews in 2006 and 2007, the research team conducted in-depth, semistructured interviews with key decisionmakers in six major Sonoran urban areas, including Hermosillo, Guaymas, Caborca, Alamos, Cd. Obregón, Navojoa, and the "twin" border cities of Nogales, Sonora, and Nogales, Arizona. In addition, we participated in a 2004 water study conducted in the cities of Naco and Cananea, led by Dr. Robert Varady and Dr. Anne Browning-Aiken at the Udall Center for Studies in Public Policy. Interview participants included urban water managers, academic experts, and environmental group leaders in each area. The river basin council study was also funded by a NOAA program housed at the Udall Center for Studies in Public Policy—National Oceanic and Atmospheric Association (NOAA) Office of Global Programs (OGP), 2003–2006, Robert G. Varady, Principal Investigator. "Use of Climate-Information Products by Water Managers and Other Stakeholders in Two GCIP/GAPP Watersheds in Arizona/Sonora and Oklahoma." The research team, led by Margaret Wilder and Nicolás Pineda Pablos at El Colegio de Sonora (Hermosillo), conducted semistructured interviews with 35 participants, including local water managers, water and irrigation officials, CNA officials, academic water experts, members of river basin councils, and environmental groups. In addition, team members attended a limited number of river basin council meetings. The irrigation districts study was conducted between 1999 and 2001 by Margaret Wilder in the Rio Yaqui Irrigation District (041) and the Altar-Pitiquito-Caborca Irrigation District (037). The study involved interviews with approximately 150 interview subjects, including CNA officials, irrigation district staff and elected officers, leaders of producers' and ejido unions, individual producers, extension specialists, and academic experts. The research was funded by an Inter-American Foundation International Dissertation Fieldwork Fellowship.

8. Some scholars take issue with the term "decentralization," claiming that it is indicative of a wholesale transfer of responsibility from federal to state or local levels, and as such, is a misnomer in cases in which the federal government retains broad responsibility to itself. Although the federal government retains policy and oversight functions, and retains some key areas of authority to itself under "decentralized" structures, the responsibility for operation and maintenance of the delivery systems nevertheless rests with the municipal or state-level authority; however, states and municipalities have limited taxation authority and do rely on federal revenue-sharing.

9. There are three primary levels of river basin (or watershed) council. At the macro-watershed (*macrocuenca*) level, are the 25 major river basin councils (or *consejos de cuenca*). At the sub-watershed level are the associated river basin commissions (or *comisiones de cuenca*), and finally, at the micro-watershed level are the river basin committees (or *comités de cuenca*) (SEMARNAP 1998, 25–26). In addition, the law establishes an integrated, user-based approach for dealing with overdrafted groundwater aquifers, called Technical Groundwater Committees (or *Comités Técnicos de Aguas Subterráneas, COTAS*).

10. The *Alto Noroeste* (Upper Northwest) *consejo de cuenca* was formally established March 19, 1999; the *Rios Yaqui-Matape consejo de cuenca,* August 30, 2000; and the Rio Mayo *consejo de cuenca,* August 30, 2000 (CNA, 2003, 79).

11. The Colonias Yaquis Irrigation District, in the traditional Yaqui Indian pueblo area just north of Ciudad Obregon, has never been transferred, although CNA reports still cite this as a goal.

12. For example, the Rio Yaqui Irrigation District of 220,000 hectares and nearly 20,000 water users has 42 separate water modules.

References

Bakker, K. 2005. "Neoliberalizing Nature? Market Environmentalism in Water Supply in England and Wales." *Annals of the Association of American Geographers* 95 (3): 542–565.

Becerril, Andrea Alma, E. Muñoz, and Carlos Camacho. 1999. "Declaran a Aguascalientes Zona de Desastre Debido a la 'Sequia.'" *La Jornada (México D.F.),* June 13.

Bauer, C. J. 1997. *Siren Song: Chilean Water Law as a Model for International Reform.* Washington, D.C.: Resources for the Future.

Búrquez, A., and A. Martinez Yrizar. 2000. "El Desarrollo Económico y la Conservación de los Recursos." In I. Almada Bay, ed., *Sonora 2000 A Debate: Problemas y Soluciones, Riesgos y Oportunidades.* Mexico D.F.: Ed. Cal y Arena.

CNA (Comisión Nacional de Agua). 2001. *Plan Nacional Hidráulico, 2001–2006.* México, D.F.: National Water Commission.

CNA (Comisión Nacional del Agua). 2003. *Estadísticas del Agua en México.* Mexico, D.F.: National Water Commission.

Cornelius, W., and David Myhre, eds. 1998. *The Transformation of Rural Mexico: Reforming the Ejido Sector.* La Jolla, CA: Center for U.S.-Mexican Studies, University of California, San Diego.

Easter, William, Herve Plusquellec, and Ashok Subramanian. 1998. "Irrigation Improvement Strategy Review: A Review of Bankwide Experience Based on Selected 'New Style' Projects." Washington, DC: World Bank, pp. 1–25.

ECLAC (Economic Council on Latin America and the Caribbean). 2007. *Social Panorama of Latin America, 2007.* Santiago de Chile: CEPAL.

Gleick, Peter H., Gary Wolff, Elizabeth L. Chalecki, and Rachel Reyes. 2002. *The New Economy of Water: The Risks and Benefits of Globalization and Privatization of Fresh Water.* Oakland: Pacific Institute.

Huber, Evelyne, and Fred Solt. 2004. "Successes and Failures of Neoliberalism." *Latin American Research Review* 29, no. 3:150–164.

Ingram, Helen, Joachim Blatter, and Pamela Doughman. 2001. "Emerging Approaches to Comprehend Changing Global Contexts." In J. Blatter and H. Ingram, eds., *Reflections on Water: New Approaches to Transboundary Conflicts and Cooperation.* Cambridge, MA: MIT Press, pp. 3–29.

Larson, Anne, and Jesse Ribot. 2004. "Democratic Decentralisation through a Natural Resource Lens: An Introduction." *European Journal of Development* 16, no. 1 (March): 1–25.

Lemos, Maria Carmen, and Antonio Luna. 1999. "BECC and Public Participation in the U.S.-Mexico Border: Lessons from Ambos Nogales." *Journal of Borderlands Studies* 14, no. 1 (Spring).

Meynen, Wicky, and Martin Doornbos. 2004. "Decentralising Natural Resource Management: A Recipe for Sustainability and Equity?" *European Journal of Development Research* 16, no. 1 (Spring): 235–254.

Pineda Pablos, Nicolás, ed. 2006. *La Búsqueda de la Tarifa Justa: El Cobro de Los Servicios de Agua Potable y Alcantarillado en México.* Hermosillo, Sonora: Colegio de Sonora.

Pineda Pablos, Nicolás. 1999. Urban Water Policy in Mexico: Municipalization and Privatization of Water Services. PhD diss., Department of Latin American Studies, University of Texas at Austin.

Prasad, Krishna C., Barbara van Koppen, and Kenneth Strzepek. 2006. "Equity and Productivity Assessments in the Olifants River Basin, South Africa." *Natural Resources Forum* 30:63–75.

SEMARNAP (Secretaría de Medio Ambiente, Recursos Naturales y Pesca). 1998. *Los Consejos de Cuenca en México, Definiciones y Alcances.* Unidad de Programas Rurales Participación Social, Coordinación de Consejos de Cuenca. Mexico, D.F.: Comisión Nacional del Agua.

Swyngedouw, Erik. 2004. *Social Power and the Urbanization of Water: Flows of Power.* Oxford: Oxford University Press.

Thobani, M. 1997. Formal Water Markets: Why, When, and How to Introduce Tradable Water Rights. *The World Bank Research Observer* 12, no. 2.

Tisdell, John G. 2003. "Equity and Social Justice in Water Doctrines." *Social Justice Research* 16, no. 4 (December): 401–416.

United Nations. 2005. "Millennium Development Goals." http://www.itu.int/ITU-D/ict/mdg/.

Varady, Robert G., D. Colnic, Robert Merideth, and Terrence Sprouse. 1997. "The U.S.-Mexican Border Environment Cooperation Commission: Collected Perspectives on the First Two Years." *Journal of Borderlands Studies* 11, no. 2:89–113.

Walton, Michael, 2004. "Neoliberalism in Latin America." *Latin American Research Review* 29, no. 3:165–183.

West, Robert C. 1993. *Sonora: Its Geographical Personality.* Austin: University of Texas Press.

Whiteford, S., and R. Melville. 2002. *Protecting a Sacred Gift: Water and Social Change in Mexico.* U.S.-Mexico Contemporary Perspectives 19. La Jolla: Center for U.S.-Mexican Studies, University of California, San Diego.

Wilder, Margaret. 2000. "The 'New Culture' of Water and the Communal Farmers of the Yaqui Valley, Sonora." *Estudios Sociales* X, no. 19 (January–June): 63–97.

Wilder, Margaret. 2002. In Name Only: Ejidatarios, Water Policy, and the State in Northern Mexico. Ph.D. dissertation. Department of Geography and Regional Development, University of Arizona.

Wilder, Margaret. 2005. "Water, Power, and Social Transformation: Neoliberal Reforms in Mexico." *VertigO: La revue électronique en sciences de l'environnement* 6, no. 2 (September). Montreal: Université de Québec de Montreal.

Wilder, Margaret. Submitted. "Hidden Practices of Privatization: A Political Ecology of Changing Access to Land and Water in Mexico's Ejido Sector."

Wilder, Margaret, and Patricia Romero Lankao. 2006. "Paradoxes of Decentralization: Neoliberal Reforms and Water Institutions in Mexico." *World Development* 34, no. 11 (November).

Wilder, Margaret, and Scott Whiteford. 2006. "Flowing Uphill Toward Money: Groundwater Management and Ejidal Producers in Mexico's Free Trade Environment." In Laura Randall, ed., *Changing Structure of Mexico: Political, Social, and Economic Prospects* New York: M. E. Sharpe, 341–358.

Wilder, Margaret, Robert G. Varady, Nicolás Pineda Pablos, Anne Browning-Aiken, Rolando E. Díaz Caravantes, and Gregg M. Garfin. In preparation. "Water Management Institutions in Mexico's 'New Culture of Water': Emerging Opportunities and Challenges for Climate Science and Climate Knowledge."

5

From Equitable Utilization to Sustainable Development: Advancing Equity in U.S.-Mexico Border Water Management

Stephen P. Mumme

The U.S.-Mexican border region, yesterday and today, has been and is constituted by water. Water frames and forms two-thirds of the boundary that divides our two nations. Twelve hundred miles of the Rio Grande and 22 miles of the Colorado compose our 1957 mile international boundary line. Water in this arid region is the staff of life, the critical element driving the region's settlement and growth. From the early expeditions of Juan de Onate to the present day, the availability of water has defined the trajectory of regional development. From Hermosillo to Phoenix water remains a critical limit on the border's human possibilities. The harnessing of water on the Rio Grande, the Colorado, and their tributaries remains among our most stupendous regional accomplishments, underwriting the growth of the region's great cities and supporting an irrigated agricultural base that is the envy of the world. Small wonder, then, that water availability and use in the border area has been a crucial test of binational relations for more than a century and that the problem of equity, or fairness, in its access and use has loomed so large historically and in our current concerns.

In matters of water governance, consideration of equity or fairness in access and distribution is one of the cardinal principles underlying every enduring water management system. This truism applies to common property issues and holds regardless of political scale. It holds equally for small associations of local appropriators and for sets of nations sharing a common watercourse. Among sovereign nations, however, the achievement of equity in managing shared waters is seldom a function of altruism; it is usually the outcome of coercion or bargaining, or some mix of each. Historically, the centrality of this vital resource has proven to be a powerful incentive to exclude neighboring states from resource access in the absence of countervailing pressures.

This is certainly true in U.S.-Mexican relations. Along the U.S.-Mexico border, the effort to fashion an equitable arrangement for appropriating and utilizing scarce water resources is as old and enduring as our sovereign relations. It is truly an epic history, full of claims and counterclaims, assertions, rebuttals, dealing and deals, bargains, speculations, and thievery, even treachery. Equity, whether conceived as altruism, or framed in the diplomatic discourse of comity, is seldom factored in where water is at stake. This characterization is particularly apt in describing binational water relations for nearly a century following the historic treaty of Guadalupe Hidalgo that cemented the border in 1848. By the mid-twentieth century, legal notions of equity began to enter into the design and formalization of water allocation practices on the principal international rivers. These norms and practices were narrowly tailored to traditional development objectives built around the commodification of water and the exclusion of significant interests in water management.

Such norms and practices are at least modestly challenged today as our nations and societies move gradually towards more comprehensive watershed management practices and more inclusionary visions of the stakes and stakeholders in border water management. These trends offer hope that equity in its broader sense may become a stronger component of management and decision making on water resources in the border area. But we are not there yet.

This chapter traces the history of U.S.-Mexican water relations to draw out the framing of equity in binational water management and to show how this has changed and is changing today. The first section looks at equity considerations in institutional design and diplomacy during the late nineteenth and early twentieth centuries. The second section describes the shift towards norms of equitable utilization and international legal prescription associated with the landmark 1944 Water Treaty. The third and fourth sections look at criticism of that normative regime and pressures for reform since the La Paz Agreement of 1983 that have led to a discursive shift towards watershed management and sustainable use of shared waters. The fifth section, a review of the All-American Canal dispute, draws attention to continuing inequity in binational water management. The chapter's final section offers some concluding observations on current challenges in broadening and strengthening the role of equitable norms in managing binational water resources.

The Gospel of Imposition: Boundary Waters and Sovereign Claims from the Treaty of Guadalupe Hidalgo through the 1906 Convention

The landmark Treaty of Guadalupe Hidalgo concluding the American-Mexican war of 1846–1848 is of interest to historians for many reasons, but particularly for its provisions fixing the land boundary between Mexico and the United States and settling most outstanding territorial claims. Coupled to U.S. envoy James Gadsden's 1853 acquisition of Mexican lands south of the Gila River for the United States, the Treaty set the territorial boundary nearly in its present lines except for numerous minor fluctuations arising from the erratic course of the Rio Grande/Rio Bravo River. What the Treaty didn't manage, however, was a division of water resources between the two countries (Hundley 1966, 18).

Whatever the reasons for this oversight, and a significant reason is simply profound ignorance of available resources in both countries at the mid-nineteenth century, the inevitable result was a pattern of unrelenting national competition over this scarce resource. The agreements of 1848 and 1853 had set a boundary that followed the middle of the Rio Grande/Rio Bravo to a point just north of El Paso, Texas, a distance of nearly 1,200 miles, which then veered overland and westward to the Pacific for some 770 miles. Along the boundary's riparian reach the Rio Grande was fed and renewed by numerous tributaries in both countries. In its overland stretch, in addition to the mighty Colorado River, the boundary bisected hundreds of arroyos, washes, and ephemeral streams as well as nearly a dozen perennial streams and rivers across the Chihuahuan and Sonoran deserts, the lower Colorado River valley, and the California coastal range. Such streams, complemented by numerous springs and *tanques* near the international boundary, were the staff of life for indigenous peoples, Mexican communities, and settlers drawn to the border after 1848 (Dunbier 1968).

While a few nineteenth-century visionaries on both sides of the international line may have imagined the desirability of achieving particular settlements allocating the water in border streams, the issue failed to draw governmental attention until late in the nineteenth century. The need for a settlement was catalyzed then by upstream diversions that diminished flows to downstream communities and was animated by rising interest generally in the potentialities and promise of irrigated lands and government sponsored reclamation. Even then, at a time when codified international law was yet in its infancy, nationalist attitudes provided few incentives for incorporating equitable considerations in diplomatic approaches to the question. Sovereign assertions and claims dominated national and local social construction of water rights and entitlements. The only handle the Treaty of Guadalupe Hidalgo provided for advancing and settling these various claims and counterclaims was found in its Articles VI and VII provisions upholding each country's right to navigation on the Rio Grande and Colorado rivers (Treaty of Guadalupe Hidalgo 1848).

Evidence of this may be seen in the earliest binational initiative to allocate a shared water asset, the protracted effort to deal with the situation on the upper Rio Grande from 1890 to 1906. By the 1870s rising Rio Grande diversions in Colorado and New Mexico had diminished flows to downstream El Paso and Cuidad Juarez, and the valley below. Local irrigators worried, but little was done until drought accentuated growing water scarcity. In 1879 Texans in El Paso pressed the State Department to approach Mexico concerning alleged illegal diversions by Mexicans in the Juarez Valley. Mexico rejected these claims. Things stood as they were until 1888 when drought again aggravated local tensions. El Pasoans commissioned a report by Anson Mills, a military engineer. On the strength of Mills' report, the U.S. Congress authorized initiation of diplomatic talks with Mexico to reach a settlement of the water dispute (Hundley 1966, 22).

At this point Mexico took the initiative, attributing the shortage to U.S. diversions in the Rio Grande headwaters (Enriquez Coyro 1975, 100). Talks stalled for various reasons. In 1895 Mexico intensified its efforts, claiming 35 million dollars in damages and accusing the United States of failing to honor the navigation clauses of the Treaty of Guadalupe Hidalgo and the Gadsen Treaty (Hundley 1966, 22).

The upshot of Mexico's diplomatic push was the now infamous determination by the U.S. Attorney General, Judson Harmon, which was soon to become known worldwide as the *Harmon Doctrine*. Harmon's reading of the treaties asserted no obligation whatsoever on the part of the United States to negotiate with Mexico over water. On the question of navigation, Harmon, disregarding the inclusive language of the original treaties and focusing on the Boundary Convention of 1884 which limited common navigation rights to the "actually navigable channels" (Timm 1941, 202), simply asserted that the Rio Grande was historically unnavigable at El Paso and that the navigation clauses only applied to the international reach of the river (Hundle, 1966). Under this reading any U.S. diversions upstream were unencumbered by its commitments on the international reach of the river. Harmon went on to sharpen the point, finding that the United States, on the basis of its absolute municipal sovereignty in international law and as upper riparian, had no obligation whatsoever to consider the impact of its water uses on Mexico (Enriquez Coyro 1975, 123). International law, he argued, required nothing of the United States, though it might consider some arrangement on the basis of "comity" (Hundley 1966, 23–24).

Despite this draconian reading, the State Department pursued talks with Mexico motivated by the urgency of the local situation. In 1896 the fledgling International Boundary Commission, its U.S. Section chaired by Mills, was asked to consider the possibility of an "equitable" division of water between the two countries. Mexico agreed. The IBC's final report attributed the problem of local scarcity to upstream diversions in the United States and supported Mexico's claim of wrongful deprivation. It called for construction of an international dam just upstream of El Paso to secure to each nation its "legal and equitable rights" with the impounded waters to be divided equally between the two countries (Hundley 1966, 24–25).

Implementation of the IBC's recommendations, unfortunately, was shortly after obstructed by the speculative efforts of the Rio Grande Dam and Irrigation Company, a British-American partnership of water speculators bent on construction of a dam nearly 125 miles upstream at Elephant Butte, New Mexico. Dealing with the U.S. Interior Department instead of State, and pursuing their case through the courts, the RGDIC's investors successfully opposed the El Paso dam, blocking the proposal until 1903 when their claims were struck down. By then, however, New Mexico state officials managed to persuade the State Department that a dam at Elephant Butte was desirable and secured congressional funds for the job. No consideration of Mexico's position was given. When Mexico protested, the United States denied it had any obligation to its downstream neighbor in international law (Enriquez Coyro 1975, 244–245). As incongruous as it seems today, the United States nevertheless offered to reach an international agreement with Mexico, amicably and on the basis of "high principles of equity." Mexico, with few options in the face of U.S. unilateralism, eventually acquiesced.[1]

The outcome, in 1906, was the U.S.-Mexico Convention dividing the waters of the upper Rio Grande (Convention 1906). By terms of this arrangement, the United States claimed more than 90 percent of an annual average flow of one million acre feet (as measured at New Mexico's Otowi gauge) on the upper Rio Grande, with Mexico receiving a guarantee of 60,000 acre-feet based on maximum known uses of Mexican settlers in the Valle de Juarez (Convention 1906; Reynolds 1968, 56). The United States agreed to build Elephant Butte dam at its expense. Mexico agreed to waive all damage claims and renounce any further claims on this reach of the river (Convention 1906). Later on, this treaty would be uniformly reviled and condemned by Mexicans, including government officials and diplomats (see, for instance, Enriquez Coyro 1975, 267, 528), who still believe Mexico was unfairly and unjustly forced to surrender its prospects for future development in the Valle de Juarez due to the sovereign imposition of the United States. Coupled with U.S. rejection of Mexico's claims to a portion of the Chamizal tract in the same region in 1911 (Lamborn and Mumme 1988), the 1906 Convention put a chill on bilateral water diplomacy that would endure for thirty years and, in truth, still taints our bilateral relations.

The Emergent Doctrine of Equitable Utilization and the 1944 Water Treaty

In the annals of bilateral water affairs and the international law bearing on the management of shared watercourses, the Harmon doctrine and the 1906 Convention stand as high-water marks of sovereign unilateralism. With historical hindsight it is clear that neither the Doctrine nor the Convention satisfied even minimal accommodations to equitable treatment, all the high-minded rhetoric to the contrary notwithstanding. So much is this so that much of the work on water management undertaken by international lawyers in the decades that followed centered on defining positions and refining arguments that would undo the damage Harmon's judgment wrought. Despite rising demand for binational accords on the lower Rio Grande and the lower Colorado by anxious settlers and developers, these turn-of-the-century events and the turbulence of the Mexican revolution stalled further cooperation until the 1920s. When in 1910 the United States sought Mexico's concession on a division of water on the lower Rio Grande, Mexico threw the Harmon Doctrine back in its face claiming absolute ownership of its tributary waters (Hundley 1966, 39). The United States, for its part, proceeded unilaterally to plan the development of the Colorado River's water resources, negotiating the

Colorado River Compact in 1922, legislating the Boulder Canyon Act in 1928, and leaving Mexico on the sidelines (Hundley 1966, 48–64). Mexico behaved similarly on the Rio Grande, moving forward unilaterally to develop the Rio Conchos and other tributaries (Enriquez Coyro 1975, 410–411).

When discussions resumed in 1924, well after the Mexican Revolution, they centered on the problem of the lower Rio Grande. These discussions proceeded in an atmosphere of profound distrust and made little progress. In 1928, alarmed by developments on the Colorado River, Mexico approached the United States with the proposal of constituting an International Water Commission to consider the disposition of international waters borderwide (Enriquez Coyro 1975, 484). The United States, anticipating Mexican objections and concerned with developments on the Rio Grande, agreed, and in 1929 negotiations on both major rivers as well as the Tijuana began (Timm 1941, 197). In discussions on the Colorado River, the United States sought to limit Mexico's claims to existing diversions, offering no allowance for future growth. Mexico, recalling the adverse terms of the 1906 Convention, rejected the U.S. position. When negotiators subsequently turned to the question of dividing water on the Rio Grande, Mexico took a similarly hard line. The IWC concluded its work in 1930, its efforts an utter failure (Hundley 1966, 73–74).

By the time the two countries were caught in the vise of the Great Depression, forty years after Harmon's verdict, little progress had been made toward reaching a binational agreement on what, in fact, represented an equitable approach to managing binational waters. The United States stood firmly by the Harmon decision and its assertion of the unilateral right and rule of the upstream riparian (Enriquez Coyro 1975, 512–513). Mexico, while not officially embracing the Harmon approach and preferring a commonwealth, or shared ownership perspective on the rivers, took the view that turnabout was fair play, asserting its sovereignty on the Rio Grande (Enriquez Coyro 1975, 512–513). Both countries were wary of utilizing the domestic rule of prior appropriation in international disputes, though the United States came to press for this on the Colorado River.

Even so, circumstances were not static. On the Canada border, the United States found reason to rue the Harmon opinion in the case of the Milk River, which originates in Canada. Bilateral discussions of Canada-U.S. boundary waters ensued, resulting in the landmark 1909 Boundary Waters Treaty between the United States and Great Britain. While

the 1909 Treaty partially incorporated Harmon's perspective in upholding each nation's right to control tributary waters of boundary rivers
and streams, it qualified this right by providing that any diversions in
one country that injured a party or parties in the other country entitled
the injured party the same right to sue as if the injury had occurred in
the country in which the diversion was made. It went on to establish the
International Joint Commission and endow it with significant powers
over boundary waters (Hirt this volume; Utton 1991a, 58–59). While
the United States had not conceded its sovereign assertion in principle,
in practice it had committed to a different, more equitable approach that
placed each nation on the Canadian-U.S. border on a common procedural
footing with reference to shared boundary waters and their tributaries;
it had, in effect, accepted the principle of limited territorial sovereignty
(Lipper 1967, 26; Meyers 1967, 570).[2]

This, then, frames the backdrop to the treaty discussions concerning a division of boundary waters that led to the landmark U.S.-Mexico
Water Treaty of 1944 apportioning the waters of the Colorado and the
middle-lower Rio Grande. As is well known, the 1944 Treaty negotiation
entailed Mexican concessions on the Rio Grande in exchange for U.S.
concessions on the Colorado River. As they approached each situation,
both countries had to face two fundamental questions: first, how should
the water be divided? both in absolute terms based on normal expectations of availability and under conditions of scarcity; second, how should
the costs of the works necessary to implement the agreement be distributed? In answering these questions the governments of both countries effectively abandoned any lingering inclination to claim absolute territorial
sovereignty and proceeded pragmatically, acknowledging the reality of
reciprocal sovereignty and informed by the growing international practice of settling water disputes on the basis of equitable apportionment
(Lipper 1967, 26–27). A detailed accounting of these negotiations has
been given in government documents and the two fine histories on the
subject by American historian Norris Hundley (1966) and Mexico's chief
negotiator in the treaty process, Ernesto Enriquez Coyro (1975). Thus,
there is no need to recapitulate this lengthy and complex story except to
draw out several important aspects that bear directly on the framing of
equity in managing U.S.-Mexican binational waters.

First, while the 1944 Water Treaty represents a bilateral shift towards
acceptance of limited territorial sovereignty as a principle of international
law (Utton 1991a, 9–11), it did not categorically embrace the principle of

equitable apportionment as had the U.S. Supreme Court in domestic interstate rivers disputes, nor did it expressly invoke anything beyond customary international law as legal precept (Lipper 1967, 26; Utton 1991a, 29–30). While this is certainly unexceptional in international treaties of this sort, it provided ample room for maneuver by either nation in dealing with subsequent matters either poorly defined by the Treaty or beyond its scope.

Second, the Treaty failed to insure the equitable resolution of a number of binational water problems, the quality of boundary waters principal among them. The problem of water quality, which received considerable attention in the negotiations, was finessed in such a way that Mexico believed it had sufficient assurances while the United States thought it gave none (Hundley 1966, 155–159). The treaty also failed to address the division of water in a number of nontributary rivers and streams crossing the border, including the Tijuana River, failed to address groundwater, and failed to make adequate provision for the protection of ecological values.[3]

Third, the Treaty embraces procedural equity within rather narrow boundaries at the government-to-government level, making no specific provision for public participation in the functioning of the agency established to give it effect, the International Boundary and Water Commission (Treaty 1944, Articles 2 and 24).

In sum, by 1945, nearly a century after the Treaty of Guadalupe Hidalgo, the two countries embraced a more equitable and cooperative approach to the management of shared water resources. This approach was limited, however, both in scope and in equity. The solution given in 1944 was neither comprehensive nor adequate to cooperatively managing a range of problems, some of which, admittedly, were difficult to envision at the time. The prevailing notion of equity contained in the Treaty was substantively bound up with sovereignty and focused on the task of apportioning the water supply of the two major boundary rivers. The Treaty was centered on, as Henry Vaux once put it, "securing national endowments" (Vaux 2000) and harnessing these to development of irrigated agriculture, municipal and industrial expansion, and hydropower. Procedurally, equity was expressed through the formalities of sovereign autonomy and diplomatic equality and given operational effect through the International Boundary and Water Commission (IBWC), whose two national sections were independent of each other and answerable only to their respective governments in applying and interpreting boundary and water agreements. Only in one small aspect did the 1944 treaty recognize the socioeconomic and administrative asymmetry that had defined the border since the late nineteenth century: Article 3 provides for subsidizing the development of binational sanitation and sewage facilities to protect U.S. border communities (Treaty 1944, Article 3).

Asymmetrical Equity: The Salinity Crisis, 1961–1973

By securing the water rights of each nation, the 1944 Treaty unleashed a surge of water development on both sides of the Rio Grande, and in U.S. and Mexican irrigation districts spread along the lower Colorado River. In this milieu, the Treaty's limits as an equitable solution to the utilization of boundary waters were soon evident. In 1961, the Wellton-Mohawk irrigation district in southwestern Arizona, acting to reduce saline groundwater accumulation in district soils with support from the U.S. Bureau of Reclamation, initiated a pumping program that discharged brackish water via the Gila River to the Colorado River mainstem at a point below Laguna dam, the last U.S. reservoir on the river. The result was an immediate spike in the salinity of Mexican treaty water and the deterioration of agricultural production in Distrito de Riego no. 14 in the Mexicali Valley (Ward 2003, 44).

The decision to dump brackish water into Mexico's treaty allotment and count it against Mexico's entitlement, while something less than an ambush, was nonetheless deliberate, cynical, and grounded in the Bureau of Reclamation's reading of the treaty's Article 10 and Article 11 dealing with Mexico's allocation on the Colorado River. During the treaty negotiations, after strenuous debate, U.S. diplomats had persuaded their Mexican colleagues to accept wording in Articles 10 and 11 that respectively read,

. . . of the waters of the Colorado River, *from any and all sources,* there are allotted to Mexico: (a) a guaranteed annual quantity of 1,500,000 acre feet. . . (Treaty 1944, Article 10, emphasis added)

and

. . . the United States shall deliver all waters allotted to Mexico wherever these waters may arrive in the bed of the limitrophe section of the Colorado River, with the exceptions hereinafter provided. *Such waters shall be made up of the waters of the said river, whatever their origin,* subject to the provisions of the following paragraphs of this Article." (Treaty 1944, Article 11, emphasis added)

This language was sufficiently ambiguous to allow each nation's negotiating team to successfully represent its own interpretation of the text to their respective senates in the ratification process (Secretaria de Relaciones Exteriores 1947, 81; Hundley 1966, 155–159). Mexican negotiators read the text in the context of the treaty and prevailing international law to ensure that its treaty water was of sufficient quality to sustain irrigated agriculture. U.S. diplomats, officials at the U.S. Bureau of Reclamation, and U.S. irrigation districts read it as blanket license to meet the Mexican quota with all available return flows to the river's mainstem.

Mexico's official protest in 1961, initiated over a decade of complex diplomatic maneuvering, aimed at securing each nation's reading of the treaty. Early in this process the foreign ministries, including the U.S. State Department, clearly understood that this was an issue of equity and that any long-term resolution of the dispute should satisfy international scrutiny on the basis of equity. Recognizing this, the U.S. position focused heavily on procedural aspects, arguing that the Treaty was fairly and mutually agreed on and that Mexico had had ample opportunity to state its views in the process of treaty ratification. As they had argued in the past, U.S. negotiators held to the view that the text of Articles 10 and 11 could not possibly be read to disallow the utilization of return flows (Zamora 1971, 4). Mexico, in response, focused on substance. It argued that the phrase "whatever their origin," however inclusive its connotative reach, did not stipulate the utilization of patently unusable water, that other articles in the treaty clearly intended that the water delivered to Mexico was to be serviceable for agricultural and domestic uses, and that international law and practice did not support the U.S. reading of the Treaty (Secretaria de Relaciones Exteriores 1975, 14; Ward 2003, 74; Zamora 1971, 13).

With the exception of officials at the U.S. Bureau of Reclamation, U.S. diplomats recognized the weakness of their position as early as 1964 as they tried to work out an interim solution.[4] The sharp deterioration of Mexican agriculture had led to strident and sustained political protests in the Mexicali Valley, attracting the press and propelling the issue to regional and national attention. The IBWC's U.S. Section and the State Department, while holding to their official position, tacitly acknowledged the merits of the Mexican case in working out an interim five-year solution in 1965. The result was Minute 218, by which the United States agreed to build at its expense and operate a bypass drain to a point just below Morelos Dam, Mexico's diversion dam for Colorado River water, to prevent further contamination of Mexican treaty waters while search-ing for a permanent solution to the problem (IBWC 1965).

In the face of intense domestic political pressure from Arizona's con-gressional delegation and other Colorado River interests, the State De-partment, though anxious to satisfy domestic interests, found itself in the uncomfortable position of advancing the importance of international law. Internal and interagency correspondence from the mid-1960's shows that Department officials were seriously concerned should Mexico elect to pursue its case through international courts (Mann 1964, 2–3; Sohn 1970; Bevans 1971). These worries deepened as Minute 218 was about to lapse in 1970. Despite the bypass, which diverted the worst of the brine from the Colorado River, salinity from district return flows, and upstream irrigation returns to the river remained severe, affecting the Mexicali Valley. In his 1970 campaign, and then as Mexico's new presi-dent, Luis Echeverria placed a salinity deal at the top of his binational agenda, intensifying Mexican pressure on the U.S. government and hint-ing it would take the question to the World Court should the United States resist.

The problem confronting the United States is nowhere better framed than in a confidential memorandum written in August 1971 by the State Department legal advisor, Steve Zamora. Reviewing the treaty negotia-tions and the legislative history in both countries, Zamora notes that while a majority of members of the U.S. Senate's Committee on Foreign Relations appeared to believe there was no water quality standard given in the treaty, that "such an unequivocal view cannot be supported by the treaty language, nor by the negotiations, nor by the conflicting testimony presented above" (Zamora 1971, 7). In Zamora's stated view, "the Com-mittee seems to be 'wishing away' the problem" (Zamora 1971, 7). He concluded with the following remarkable assessment:

. . . it is my belief, based on the examination of materials outlined above, that a minimum water quality standard "water suitable for irrigation purposes" was a fundamental assumption underlying the Water Treaty of 1944. This means not only that the United States cannot act unreasonably in its use of the water upriver, but also that, under the doctrine of *rebus sic stantibus,* if the situation ever arises in which a *substantial part* of the water delivered to Mexico is of unusable qual-ity, Mexico may terminate or withdraw from the Treaty or suspend its operation. (Zamora 1971, 17)

In short, the Zamora memorandum supported the idea that a minimum standard of substantive equity was to be found in the treaty record and that the United States could not ignore this downstream obligation. Zamora's opinion suggested a losing case for the United States at the World Court.

On the basis of this and other opinions, the State Department successfully pressed U.S. interests for an accommodation while working all the while to prevent Mexico from taking its case to an international forum. Responding to Mexican initiative and a buildup of pumping in the Mexicali Valley, it also seized the opportunity to advance the importance of considering groundwater management at the binational level, addressing another important lacuna in the treaty. The eventual solution, Minute 242, signed in August 1973, committed the United States to deliver to Mexico water of equivalent quality to that impounded at Imperial Dam, the lowest U.S. storage dam on the river, provided for an extended bypass drain for Wellton-Mohawk brine, limited groundwater pumping on the San Luis mesa, and required binational consultation in advance of any domestic action that would affect groundwater utilization in the neighboring country (IBWC 1973). It clearly vindicated Mexico's original complaint and is fully in the spirit of the concerns laid out in Zamora's memorandum.

In many respects, the salinity crisis represents a transition in bilateral deliberations on water and the role of equity in these deliberations. Not only did it compel the two countries, and particularly the United States, to consider questions that were not directly addressed in the 1944 Treaty and admit to a broader construction of the basis of equity in the case of water quality, but it broadened the range of substantive matters within the Treaty's reach, as seen in the case of groundwater. Yet it really did more. Minute 242 formally addresses substantive equity within the scope of the Treaty, but it also responded, importantly, to growing procedural concerns and an emerging set of stakeholders. Buried with the voluminous documentation on the conflict, one sees an awareness of differentials of wealth and poverty between the two countries, shaping some official attitudes on fairness (American Embassy 1972a). We see the emergence of environmentalists as a force to be reckoned with as State Department officials, confronted with the domestic implications of the new National Environmental Policy Act (1969), worried that nongovernmental organizations might sue the Bureau of Reclamation for acting to promote the pollution of a national and international waterway (American Consul 1972a; American Embassy 1972b). The politics of the Salinity Crisis thus mark a transition from older to newer ways of construing or defining the relationship between equity, water, and development; between equity expressed as sovereign command of resources and equity conceived in a much broader social context, and between equitable utilization as a narrowly distributive concept and equitable utilization as a qualitative concept. By the early1980s these issues would be firmly cobbled to the binational docket.

La Paz and the Emergent Ethics of Sustainability

While Minute 242 settled the salinity dispute on the lower Colorado and compelled the United States to deal with the salinity problem basin-wide, it also had the practical effect of highlighting the problem of pollution in border streams. After 1970, the salinity issue had been framed by environmental activists and academics as both a pollution and an ecological problem, with potential public health and NEPA implications (American Consul 1972a; American Embassy 1972c). By 1973, these issues were loosely on the border radar screen, attracting local policy attention to the inadequacies in the treaty framework for managing boundary water.

Public awareness of pollution, as both concept and condition, represented a reframing of a much older problem. Until 1970, contaminated water had been largely associated with sanitation and the public health of urban settlements on the border. At locations like Nogales and Tijuana, sewage contamination plagued Mexican communities, with frequent spillage in border arroyos and streams. The problem was sufficiently threatening to U.S. cities downstream that the IBWC was given the mandate of working out binational solutions to border sanitation problems. But the notion of pollution, hitched to the emerging notion of ecology, stretched the range of public concerns and policy considerations associated with water contamination and, importantly, admitted new stakeholders with regulatory interests in border water. With U.S. adoption of NEPA in 1969, and new environmental rules shortly after in Mexico in 1971, the need to consider a wider range of potential water problems at the border was established in domestic policy. The binational basis for managing pollution was less clear.[5]

This soon became evident. In 1977, a little more than four years after the Minute 242 was signed, spillage from copper tailing ponds at Cananea, Sonora reached the headwaters of the Rio San Pedro, contaminating water, killing fish, and polluting irrigation water downstream (Jamail and Ullery 1979, 37). The San Pedro River enters the United

States east of Arizona's Huachuca Mountains, passing various farms and ranches and skirting the city of Sierra Vista, Arizona on its long meander towards the Gila River.

State officials, believing the problem properly fell within the scope of the agency's treaty authority, took the matter to the IBWC in January 1978 (Jamail and Ullery 1979, 41). The U.S. Commissioner, Joe Friedkin, agreed to press for mitigation with Mexico. Despite this commitment, the U.S. Section seemed to procrastinate. Friedkin's hesitance in pressing Mexico for immediate action was understandable. Since its establishment in 1889, and certainly since its expansion in 1944, the only regulatory role the IBWC had historically discharged was boundary alignment. Its self-image and experience had been shaped as a water provider for border constituents. Friedkin, a key architect of the 1963 Chamizal Agreement, the 1970 Boundary Treaty, and Minute 242, was uncertain of his treaty authority and clearly reluctant to risk tarnishing his agency's favorable image by entering the controversial arena of environmental regulation. He wished to proceed with caution. The U.S. Section's insularity and procrastination angered environmentalists, who turned directly to the State Department and the fledgling Environmental Protection Agency's small international office. The EPA, sensing an opportunity to enhance its international mandate, joined the U.S. Section and State Department officials in placing border water pollution issues on the agenda of the February 1978 presidential summit between U.S President Jimmy Carter and Mexican President Jose Lopez-Portillo (Jamail and Ullery 1979, 44). The summit produced a joint memorandum by which Mexico agreed to deal with the problem on the San Pedro. The EPA quickly followed up by negotiating a further joint memorandum on U.S.-Mexico environmental cooperation agreement in June 1978, the first formal agreement between EPA and its Mexican counterpart (EPA 1978).

The momentum generated by the San Pedro River crisis catalyzed further binational initiatives. The EPA, acting under the authority of its new agreement, undertook an inventory of border environmental problems, drawing public attention to a range of environmental problems along the border (Hunt 1980). The IBWC, sensing its 1944 Article 3 authority under challenge, sought to shore up and amplify its treaty-based jurisdiction, signing Minute 261 a year later. Minute 261, building on Article 3, broadened the Treaty's definition of border sanitation problems to incorporate sanitary conditions that impair the beneficial use of all waters crossing the international line (IBWC 1979). By this minute, the

IBWC, for the first time, gained partial jurisdiction over all transboundary water flows. But its jurisdictional reach remained strictly confined to international water quality problems, falling well short of a mandate for other environmental issues, including those linked to water availability (Mumme 1981). And while the new minute sought to send a message that sanitation was now a higher priority on the governments' agenda, it did little to respond to public criticism of the IBWC's institutional accountability and responsiveness.

The need for a better substantive and procedural solution was evident, and with EPA leading the way, a solution was not long in coming. In August 1983, four years after Minute 261, the two countries signed the U.S.-Mexico Border Environment Cooperation Agreement, better known as the La Paz Agreement for the Mexican coastal city in which it was struck (Agreement 1983). Unlike Minute 261, which honed sharply to the substantive and procedural aspects of equity associated with the 1944 Treaty, the La Paz Agreement broadens each of these elements. On the substantive side, the Agreement frames border water pollution within the broad rubric of environment and ecology, recognizing the IBWC's unique administrative responsibility for transboundary water problems, yet broadening the scope of concerns justifying binational attention (Agreement 1983). With a broad definition of the border region, the La Paz Agreement allows consideration of problems in this international forum that might otherwise be treated as strictly domestic matters. Procedurally, the La Paz Agreement breaks new ground by officially granting standing to states, municipalities, and nongovernmental entities in binational deliberations on environmental matters and provides a forum for their participation (Agreement 1983, Article 9).

Even prior to La Paz, the social construction of equity in border water management had begun to shift from a narrow sovereign preoccupation with water endowments and command of the region's water resources toward a broader set of substantive and procedural concerns. With La Paz, this emergent ethic became more visible. The new ethos entailed a more holistic understanding of water as a resource embedded in watersheds and ecosystems. It linked quantity and quality. It integrated the consideration of water with other natural and social processes and did so in a broader spatial frame, directing public attention to social conditions on both sides of the border. This new perspective included an emphasis on public health and social awareness of place, community, and habitat. It recognized the importance of new players and participants in border water policy, valued open policy forums, and emphasized the need for official transparency. A dialogue among and between twin cities and paired communities was emerging. While this new ethos was additive and far from consolidated in the border community, it certainly didn't displace sovereign reckoning in binational water management, but it had begun to challenge older, more traditional approaches to water management. Though not yet present in border water policy dialogue and debate, one can see in La Paz the incorporation of some of the elements associated with the idea of sustainable development.

Water Management since NAFTA: The Challenge of Sustainability

For environmentalists, the decade following the La Paz Agreement was, in many respects, disappointing. The La Paz process itself directed attention to water-based environmental threats and facilitated raising these issues on the binational agenda. But the process was plodding, ad hoc, and largely dominated by traditional players in binational water policy, federal and state governments, the latter enhanced by new environmental offices, municipal utilities in the larger border cities, and irrigation districts.

NAFTA presented an opportunity to alter this equation. Environmental debate on NAFTA centered on the border community and on water. In policy circles the border was described as a veritable cesspool. Trade expansion and new investment were certain to add additional environmental stressors, generating new water demand and taxing the border's already inadequate water and sanitation infrastructure. In short, in raising the specter of greater demand, new water transfers, and water pollution in the border zone, free trade put a spotlight on the ethics of unequal development, including the development of water resources. The question of sustainability was now squarely on the table, rhetorically at least.

This new rhetoric permeates the discussion of contemporary border water problems, beginning with the Integrated Border Environmental Plan (IBEP) in 1992 (EPA 1992). It's embedded in the NAFTA institutions, the Border Environment Cooperation Commission (BECC), the North American Development Bank (NADB), and the Commission for Environmental Cooperation (CEC). With IBEP, sustainable development became a formal binational objective under the authority of the La Paz Agreement and a policy principle for border water planning carried forward in the 1996–2000 Border XXI Program and its successor, the

Border 2012 Program. Sustainability elements for environmental protection included economic development, public health, administrative decentralization, local capacity building, and public participation in policy design and implementation.[6] For water management, this has meant heightened attention to meeting basic water needs in border communities and greater emphasis on ecological values.

The upside and downside of this sustainability language is nowhere more apparent than in the 1993 charter for the functionally linked BECC and NADB organizations. Under the 1993 Agreement to Establish a Border Environment Cooperation Commission and North American Development Bank both agencies are expected to promote sustainable development in advancing improvement of the border's environmental infrastructure (Spalding 2000, 126). The BECC, with a mandate to promote the development of needed water infrastructure in the border area, is instructed to apply sustainable development criteria in certifying projects for funding, to operate in a publicly accountable and transparent manner, and to promote public participation in the development and approval of proposed projects (Agreement 1993). Projects meeting the agency's high sustainability criteria are fast-tracked for approval.

And yet, while this new sustainability emphasis draws attention to shortcomings and inequities in border area water management, it hardly entails a systematic or comprehensive commitment to socioeconomic disparity reduction at the border. Such issues were raised in the NAFTA debate but largely set aside. Evidence of this is perhaps best seen in the authority of BECC's partner agency, the NADB. While U.S. financial interests were willing to support the establishment of a border development bank, they were unwilling to subsidize its functions, restricting NADB's lending to at-market rates. The predictable outcome on both sides of the border was minimal demand for NADB's facility (Reed and Kelly 2000, 6).

If the new discourse of sustainable development largely failed in driving the governments to attend to glaring inequities in binational water allocation and management, it did spotlight important procedural aspects of environmental equity, focusing attention on a deficit of government transparency and responsiveness to local concerns. The BECC's project approval procedures were well received by environmental and community organizations focused on local infrastructure provision in the border region (Spalding 2000, 125–126). Other agencies, including the historically insular IBWC, began to upgrade public relations and citizen outreach.[7]

Under the joint Border XXI Program and, more recently, Border 2012, the U.S. EPA and Mexico's SEMARNAT have promoted participatory mechanisms, creating joint task forces and both regional and watershed advisory bodies (EPA 2000, 2002). The EPA's citizen-government advisory board, the Good Neighbor Environmental Board, actively champions public participation and government accountability, producing three reports on border water management in the last decade, each advocating partnerships and voice in border water management (see, for instance, the most recent report, GNEB 2005, 31).

There is no question that many of these new public participation venues remain more symbolic than real. While a new discourse of stakeholders, integrated watershed management, water councils, water task forces, and binational dialogue has taken hold, it remains to be seen whether these developments will have real force in shaping border water policy. Old habits of corporatist representation and social exclusion are mirrored in some of the U.S. groups recently created by various federal and binational agencies[8]. Impediments to effective participation abound in the border region where poverty and diminished social organization still restrict the public's voice. Declining funding for Border 2012, BECC-NADB, and border water programs in general since 2000 (GNEB 2005, 15) certainly does little to encourage public participation or strengthen public capacity for water policy decision making.

A related development is a new emphasis on environmental justice, presently expressed largely at the national level. Already an element of U.S. federal environmental policy, by the mid-nineties environmental justice was written into the new repertoire of post-NAFTA border programs. Communities concerned with the socially adverse impacts of project development received further legitimation and access to legal remedies. By the end of the 1990s Mexico followed suit, adopting its own environmental justice program for its northern border region.

These border-focused environmental justice programs vary significantly on either side of the line, reflecting important socioeconomic, cultural, legal, and administrative realities in each nation (Mumme 2006). Common to each, however, is a fairly narrow emphasis on public health that restricts the scope of distributive justice concerns (see chapter 3). Even so, national attention to environmental justice supplements the toolkit of local activists as they confront the adverse impacts of project development on border area water resources. At the international level, however, the failure to advance the important agenda of transboundary environmental impact assessment still hobbles any effort to hold each nation accountable for adverse environmental effects on the other country with implications for environmental justice (CEC 2003, 3). While harmed citizens of either country may litigate in the other nation's courts, greater binational cooperation is needed if environmental justice enforcement is really to span the border.

In sum, the past decade has seen improvement in the way the two nations address the basic needs for water infrastructure in poor communities along the border. There has also been a very modest advance toward more equitable use of transboundary water resources. While much of the rhetoric of sustainable development remains unrealized, the fact that a small aspect of the binational dialogue on water management and utilization is changing is certainly progress. That new public participation venues offer at least a table seat and a distant vision of sustainable water management to diverse stakeholders is a sign of progress. The fact that water-rights holders must at least exercise and defend their entitlements with modestly greater reference to the public good may also be seen as progress. Yet it is instructive that much, indeed, most of our progress in pursuing equitable uses of water resources has come through the portal of public health, not the market, nor the legal system of apportionment and rights that underwrites the market. One need not look too far along the border, or too closely, to discern resistance to sustainable development and the ecological and humane values it supports. Consider, for instance, the case of the All-American Canal.

The All-American Canal

The All-American Canal ranks among the most notorious symbols of the era of water grabs and imposition along the U.S.-Mexican border. Begun in 1939 and completed before the 1944 Water Treaty took effect, the eighty-two-mile canal paralleling the border with Mexico was meant to end U.S. dependence on Mexico's Alamo Canal and bolster California's claim to a larger share of Colorado River water in advance of the 1944 Treaty negotiations. The Canal certainly served its purpose. It remains the most important conduit of Colorado River water bound for California. And California is the largest user of the River.

In 1942, when the AAC entered fully into operation, all major parties on the lower Colorado River understood that water seeped to the Colorado River aquifer and that seepage followed the Alamo River and

AAC channels (Ward 2003, 32; Enriquez Coyro 1975, 628–630; Hundley 1966, 132–133). Indeed, the United States agreed to deliver part of Mexico's quotient of Colorado River water to the Alamo Canal via an AAC diversion south of Pilot Knob knowing full well at the time it would generate seepage (Treaty 1944, Articles 11 and 15). Groundwater seepage was part of the natural and human ecology of the Colorado River basin. Unfortunately, the 1944 Treaty failed to allocate this resource or recognize it as a potential binational issue (Treaty 1944; Enriquez Coyro 1975, 914). U.S. success in limiting Mexico's allocation on the Colorado River led directly to greater Mexican reliance on groundwater, a part of which percolated across the border from the AAC.[9] When U.S. drainage polluted Mexican water in the Salinity Crisis, 1961–1973, Mexican dependence on groundwater increased (Ward 2003, 93).[10] The solution to the Salinity Crisis adopted in 1973 for the first time formally acknowledged groundwater as a binational issue without specifically mentioning the AAC as a contributing problem (IBWC 1973).

Less than a decade after the Salinity Crisis was settled, conservation concerns in California drew attention to the desirability of recapturing an estimated 80,000 acre-feet of seepage lost to Mexico from the AAC. As early as 1981, in response to state criticism (Jones 1981, A3), the Imperial Irrigation District agreed to transfer water to other Southern California users provided the AAC was lined, recapturing water destined for Mexico for use in the United States (Waller 1992, 14–15). In the late 1980s the U.S. Congress passed legislation funding the canal's lining (Metropolitan Water District of Southern California 2002). Shortly after, as a requisite of project approval, a hastily organized environmental impact study was mounted, with formal but minimal consultation with Mexico and virtually no direct consultation with users in the Mexicali Valley (Jones, Duncan, and Mumme 1997). Even more damning, the 1994 EIS failed to account for critical biota and ecological resources in Mexico affected by lining the canal (BOR 1994).

Focused on the North American Free Trade Agreement in the early 1990s, Mexico was unwilling to make a serious issue of the problem but it did express its reservations. In the meantime, California proceeded with plans for the lining, which, following the 1994 preferred alternative, now entailed building a new lined canal parallel to the old one (Metropolitan Water District 2002). Despite the projected adverse direct impact on more than one thousand Mexican farm families (Cortez-Lara 2005, 279), and emerging ecological concerns based on new field evidence

(Snape, Acuna, and Gaxiola 2005), the United States refused to modify its plans, though the IBWC and the U.S. Bureau of Reclamation agreed to discuss possible mitigation actions as a matter of comity. In 1999, a Bureau of Reclamation brokered a conservation agreement among the Colorado River basin states aimed at bringing California into compliance with the Law of the River further reinforced the lining project (Bureau of Reclamation 2000, 14–15).

What is interesting is that at no point, despite Mexican reservations and protests, did the United States feel seriously obligated to consider Mexican concerns in project design or in allocating saved water. The U.S. position simply asserted that seepage was part of its Treaty-allotted water supply and, hence, entailed no obligation to Mexico save notification, to comply with terms of IBWC Minute 242 which settled the dispute over Colorado River salinity (Utton, 1991b; Kishel 1993). This, of course, begged the question of groundwater as related to the Treaty itself.

In the summer of 2004, a coalition of Mexican and U.S. water and environmental interests sued the U.S. Department of the Interior in U.S. Federal Court in Las Vegas, Nevada, asserting administrative irregularities in the original environmental impact assessment process, disregard for ecological values, disregard for international law, and disregard for Mexico's historic reliance on groundwater in the northern part of the Mexicali Valley occasioned, in part, by unilateral U.S. actions (Dribble 2005; Snape, Acuna, and Gaxiola 2005). Just how this will be resolved is uncertain but it has driven the Fox Administration to lodge a formal diplomatic note of protest with the United States government, abandoning Mexico's soft-glove approach to the issue. At the center of this unusual binational lawsuit, of course, lies the question of equity, and a critique of the heavy-handed unilateral approach the United States has historically taken to claiming transboundary water resources.

As a classic conflict in law with billions of dollars in potential water values at stake, this case, unfortunately, will be only slightly influenced by progressive notions of sustainable development. Its resolution is more likely to turn on questions of administrative impropriety related to the domestic enforcement of extant environmental law than on broader issues of equitable utilization or international obligation. And yet, it is hard to see the heart of the problem in terms of anything other than equity. What the AAC case tells us, at root, is that the real issue of equitable utilization is still trumped by upstream location and raw national assertion. As one prominent water attorney who will go unnamed put it, this is an issue that pits the interests of Southern California real estate magnates peddling million-dollar homes against the daily livelihoods of thousands of Mexican *campesinos*. Yet the smart money bets that the realtors are likely to win this asymmetrical dispute.

Deepening and Strengthening Equitable Management of Border Water Resources: The Contemporary Challenge

The All-American Canal seepage dispute is a sobering reminder that equity remains elusive in U.S.-Mexican water management. As a principle of binational cooperation it has gradually insinuated itself in an evolving set of norms and practices that arguably set limits on sovereignty's excesses, directs attention to the modern norms of international law, and better incorporates a broader range of actors and interests in the crafting of binational water policy. In the past decade, greater material investment has been made available to improve conditions in poor communities along the border. Yet for all that, the utilization of boundary water is still fundamentally proprietary, reinforced by sovereign assertion.

And yet, as recent history shows, this proprietary, asymmetrical model of water ownership and use is less serviceable for both nations today. Not only does it perpetuate old injustices and inequities, it diminishes our collective security, restricts our markets, and materially complicates solutions to a wide range of compelling water-related problems of interest to the border community. The prosperity and progress of border communities is heavily dependent on water resources. As border cities and communities become more tightly linked socially and economically, a deficiency of water on one side of the border is sure to affect the other. In some instances, the very hydrology of our natural watersheds directly ties the sustainable development of one community to the other. This is certainly true of groundwater and applies to many surface streams as well. An increasing appreciation of the many legitimate ecological claims on our allocated waters raises the further question of who bears the burden of reallocating water to meet these timeless needs when ecosystems are so clearly linked across international boundaries. The equity problem is complicated further by human economic and defensive interventions that disrupt natural systems, impair water resources, and create serious hazards for border residents.

Achieving greater equity in binational water management is thus vital to the stewardship of border water resources. Solutions to these many

problems require bilateral cooperation at the national level and the cooperation of tribes, states, counties, cities, and citizens linked by the international line. Bilateral cooperation at any level is affected by equity and
the perception that equity considerations, procedural and substantive,
are in play. The shift toward a contemporary framing of water utilization
in the language of sustainable development adds moral force to demands
for greater equity as a condition for bilateral cooperation in water allocation and management. The institutionalization of environmental justice
as a legal and social norm reinforces such concerns.

Embracing and deepening commitment to equity in binational water
management turns not just on progress in international law and on the
common understanding now emerging on managing international watersheds in the global community. Perhaps more importantly, it turns on
maintaining and strengthening institutional commitments to sustainable
development and building these notions into domestic and binational
practices. It means adequately funding our binational programs and
reaching new international agreements on conservation and watershed
management. It means strengthening old institutions in a manner that
harnesses these agencies to new values. It means enabling and encouraging citizens to better utilize the legal and administrative procedures available to citizens in both countries. In essence, it means civic empowerment
and commitment to a different set of development norms, norms that
appreciate the prevailing asymmetries and resource differentials at the
border and the intertwining of our economies and cultures.

In sum, advancing equity in border water management is essential for
future binational cooperation and the sustainable development of the
border region. This will happen more quickly if proprietary actors adopt
a broader and more complex view or perspective of their water security
and the many ways that water truly links and bridges the border divide.
In the final analysis, building in equity should be understood both as a
moral good in itself and as a critical element of the architecture of North
American interdependence. Where water is concerned, it is truly an investment in our common future.

Notes

1. Lipper argues that the Harmon doctrine was never definitively applied as a
matter of international law as the principle is not stated in the Treaty, which
instead refers to "comity" as its basis (Lipper 1967, 27). It may be worth not-

ing that the U.S. State Department, in dismissing Mexico's protest prior to the conclusion of the Treaty makes no direct reference to the Harmon doctrine either (Enriquez Coyro 1975, 244–245).

2. U.S. federal and state courts were already employing the rule of equitable apportionment in adjudicating domestic water disputes (Lipper 1967).

3. It is more accurate, perhaps, to say the Treaty made inadequate provision for emerging values in border water management as it consigned all unspecific concerns to the bottom of its list of beneficial water use priorities (Treaty 1944, Article 3).

4. See, for instance, the State Department's memorandum summarizing its views on the Salinity Crisis for Henry Kissinger in 1971 (Elliot 1971, 1).

5. As early as 1971, then U.S. IBWC commissioner Joe Friedkin advocated congressional funding for an IBWC-led study of pollution in the Rio Grande River. As far as I can determine, this is the first binational initiative to study pollution per se on the U.S.-Mexican border, and certainly on a region-wide basis. Impetus for this may have been the reported mercury contamination of the Rio Grande near Terlingua, Texas. The authority for such an effort, which was actually done by the Department of Health, Education, and Welfare's Public Health Service, was presumptively Article 3 of the 1944 Water Treaty (Friedkin 1973; Friedkin 1971).

6. See, for instance, the introductory language of the Border XXI framework document (EPA, 1996).

7. As an example, see the U.S. IBWC's new Strategic Plan (IBWC 2000).

8. The U.S. IBWC's Citizen Forums are certainly open to criticism in this regard. Under controversial commissioner Arturo Duran (2004–2005) the newer Citizen Forum's were composed of a questionable number of the Commissioner's personal friends and associates.

9. Mexican reliance on groundwater intensified after 1950, due in large measure, to U.S. river operations that reduced the flow of the Colorado River to Mexico from an average of 2 maf annually prior to 1954 to the Treaty quota of 1.5 maf. Mexican farmers were troubled by the reduction and responded by drilling wells to augment water supply (Ward 2003, 38). Between 1953 and 1958 Mexico's National Irrigation Commission drilled 396 wells capable of irrigating 40,000 hectares in the Mexicali Valley (Orive Alba 1970, 114). By 1964 more than 600 groundwater wells were in operation in the Mexicali Valley (Enriquez Coyro 1975, 924).

10. The Salinity Crisis, caused by drainage of saline Wellton-Mohawk irrigation water to the Colorado River quotient flowing to Mexico under the 1944 Treaty, provoked a near 50 percent increase in Mexican reliance on well water after 1962. By 1972 a total of 625 wells (225 private and 400 government-owned) were recorded in the Mexicali Valley (American Consul 1972b). Well water was critical for dilution of the highly saline Colorado River water flowing into Mexico below Yuma, Arizona. Thus, it is hard to avoid the conclusion that Mexican

dependency on groundwater in the Mexicali Valley is substantially a function of U.S. unilateral decision making that provoked the Salinity Crisis. The former Mexican Commissioner of the CILA (IBWC), Joaquin Bustamante, notes in his recent book, *La Comision Internacional de Limites y Aguas entre Mexico y los Estados Unidos,* that another reason for Mexican pumping south of the AAC was that AAC (and presumptively the Alamo Canal as well) so raised groundwater levels it posed a threat to Mexican farming in the area (Bustamante 1999, 492–493).

References

Agreement between the United States of America and the United Mexican States on Cooperation for the Protection and Improvement of the Environment in the Border Area. 1983. U.S.-Mex., TIAS 10827, August 14.

Agreement between the Government of the United States of America and the Government of the United Mexican States Concerning the Establishment of a Border Environment Cooperation Commission and a North American Development Bank. November 16. Available on the Web site of the Border Environment Cooperation Commission, www.cocef.org.

American Embassy. 1972a. Confidential Telegram from American Embassy, Mexico City, to Department of State. Subject: Salinity as a Political Issue. NACP, Record Group 59, Mex-U.S., POL 33–1.

American Embassy. 1972b. Limited Official Use Telegram from American Embassy, Mexico City to Department of State. Subject: Salinity, National Wildlife Federation. February 10. NACP, Record Group 59, Mex-U.S., POL 33–1.

American Embassy. 1972c. Limited Official Use Telegram from American Embassy, Mexico City, to Secretary of State. Subject: Salinity, Continued Press Attention. NACP, Record Group 59, Mex-U.S., POL 33–1.

American Consul, Mexicali. 1972a. Limited Official Use Airgram from American Consul, Mexicali to Department of State, February 16. Subject: Colorado Radioactivity Flap. NACP, Record Group 59, Mex-U.S., POL 33–1.

American Consul, Mexicali. 1972b. Steps to Protect Mexicali Valley Water Table. Unclassified Airgram from American Consul, Mexicali to Department of State, May 26. NACP, Record Group 59, Mex-U.S., POL 33–1.

Bevans, Charles I. 1971. Confidential memorandum from Charles I. Bevans, Office of Law and Treaties, to Mark B. Feldman, Assistant Legal Advisor, American Republics Affairs, U.S. Department of State: U.S. Commitment to Arbitrate, September 14. NACP, Record Group 59, Mex-U.S., Pol 33–1.

Bustamante, Joaquin. 1999. *La Comisión Internacional de Limites y Aguas entre México y los Estados Unidos.* Ciudad Juarez: Universidad Autónoma de Ciudad Juarez.

Bureau of Reclamation. 1994. *All-American Canal Lining Project: Final Environmental Impact Statement.* Washington, DC: BOR, Department of the Interior.

Bureau of Reclamation. 2000. *Colorado River Interim Surplus Criteria: Final Environmental Impact Statement.* Washington, DC: BOR, Department of the Interior.

CEC (Commission for Environmental Cooperation). 2003. *Operational Plan of the Commission for Environmental Cooperation, 2004–2006, Revised Draft.* Montreal (Quebec), Canada: CEC Secretariat, October 31.

Convention. 1906. Convention between the United States and Mexico Providing for the Equitable Distribution of the Waters of the Rio Grande for Irrigation Purposes. May 21, 1906, 34 Stat. 2943.

Cortez-Lara, Alfonso. 2005. "Enfoques encontrados en la gestion de recursos hidraulicos compartidos. El Revestimiento del Canal Todo Americano y el Valle de Mexicali: Equilibrio estatico de Mercado o equilibrio de Nash?" in Vicente Sanchez Munguia, ed., *El Revestimiento del Canal Todo Americano.* Tijuana: El Colegio de La Frontera Norte y Plaza y Valdez Editores, pp. 273–293.

Dribble, Sandra. 2005. "U.S.-Mexican Groups Sue Dept. of Interior Over Water," *San Diego Union-Tribune,* July 20.

Dunbier, Roger. 1968. *The Sonoran Desert: Its Geography, Economy, and People.* Tucson: University of Arizona Press.

Eliot, Theodore L. 1971. Memorandum for Mr. Henry A. Kissinger, the White House. Subject: Colorado River Salinity Problem with Mexico. Classified Confidential. April 12. U.S. National Archives, Record Group 59, Pol 33–1, Mexico-U.S.

Enriquez Coyro, Ernesto. 1975. *El Tratado entre México y los Estados Unidos de América sobre Ríos Internacionales, tomo I y tomo II.* Mexico, D.F.: Facultad de Ciencias Políticas y Sociales, Universidad Nacional Autónoma de México.

EPA (Environmental Protection Agency). 1978. Memorandum of Understanding between the Subsecretariat for Environmental Improvement of Mexico and the Environmental Protection Agency of the United States for Cooperation on Environmental Programs and Transboundary Problems. Mexico City, July 6.

EPA (Environmental Protection Agency). 1992. Integrated Border Environmental Plan for the Mexican-U.S. Border Area (First Stage, 1992–1994). Washington, DC: U.S. EPA, A92–171.

EPA (Environmental Protection Agency). 2000. U.S.-Mexico Border XXI Program: Progress Report 1996–2000. Washington, D.C.: U.S. EPA, EPA160/R/00/001.

Friedkin, Joe. 1973. Letter to Chris G. Escrow, Esq. U.S. National Archives, Record Group 59, Pol 33–1, Mexico-U.S.

Friedkin, Joe. 1971. U.S. Embassy, Mexico: Telegram. Subject: Comments on Rio Grande Contamination. March 16. U.S. National Archives, Record Group 59, Pol 33–1, Mexico-U.S.

GNEB (Good Neighbor Environmental Board). 2005. *Water Resources Management on the U.S.-Mexico Border.* Washington, DC: GNEB, Eighth Report to the President of the United States, EPA 130-R–05–001.

Hundley, Norris. 1966. *Dividing the Waters*. Berkeley, CA: University of California Press.

Hunt, Walter J. 1980. "Inventory of Pollution Problems along the U.S.-Mexico Border." Washington, DC: EPA International Office. Paper on file with the author.

IBWC. 1965. Recommendations on the Colorado River Salinity Problems. Cuidad Juarez, March 22. On file with U.S. IBWC at www.ibwc.state.gov.

IBWC. 1973. Permanent and Definitive Solution to the International Problem of the Salinity of the Colorado River, Minute 242. Mexico, D.F., August 30. On file with U.S. IBWC at www.ibwc.state.gov.

IBWC. 1979. Recommendations for the Solution to Border Sanitation Problems, El Paso, September 24. On file with U.S. IBWC at www.ibwc.state.gov.

IBWC. 2000. *Strategic Plan*. El Paso: U.S. Section, IBWC.

Jamail, Milton H., and Scott J. Ullery. 1979. *International Water Use Relations along the Sonoran Desert Borderlands*. Tucson: Office of Arid Lands Studies, University of Arizona.

Jones, Lilias C., Pamela Duncan, and Stephen P. Mumme. 1997. "Assessing Transboundary Environmental Impacts on the U.S.-Mexican and U.S.-Canadian Borders." *Journal of Borderlands Studies* 12 (Spring and Fall): 73–96.

Jones, Robert A. 1981. "Study Urges Cut in Irrigation Waste." *Los Angeles Times,* April 20, 3A.

Kishel, Jeffrey, P. E. 1993. "Lining the All-American Canal: Legal Problems and Physical Solutions," *Natural Resources Journal* 33 (Summer): 697–726.

Lamborn, Alan, and Stephen P. Mumme. 1988. *Statecraft, Domestic Politics, and Foreign Policy Making: The El Chamizal Dispute*. Boulder: Westview Press.

Lipper, Jerome. 1967. "Equitable Utilization." In A. H. Garretson, R. D. Hayton, and C. J. Olmstead, eds., *The Law of International Drainage Basins*. New York: Oceana Publications, pp. 15–87.

Mann, Thomas C. 1964. Letter from Thomas C. Mann, Assistant Secretary of State, to Secretary of the Interior, Stuart L. Udall, April 20. Classified Confidential. Tucson: Stuart L. Udall Archives, University of Arizona Special Collections, Main Library.

Metropolitan Water District of Southern California. 2002. "All-American and Coachella Canals." Available at MWDSC's Web site, www.mwdh2o.com/mwdh2o/pages/yourwater/supply/conservation/conserv01.html.

Meyers, Charles. 1967. "The Colorado Basin." In A. H. Garretson, R. D. Hayton, and C. J. Olmstead, eds., *The Law of International Drainage Basins*. New York: Oceana Publications, pp. 486–607.

Mumme, Stephen P. 2006. "Equity and Justice in Binational Environmental Policy." In Jane Clough-Riquelme and Nora Bringas Rábago, eds., *Equity and Sustainable Development: Reflections from the U.S.-Mexico Border*. Boulder, CO: Lynn Rienner, pp. 325–336.

Mumme, Stephen P. 1981. "The Background and Significance of Minute 261 of the International Boundary and Water Commission." *California Western International Law* l, no. 11 (Spring): 223–235.

Orive Alba, Adolfo. 1970. *La Irrigación en México*. Mexico: Editorial Grijalbo, S.A.

Reed, Cyrus, and Mary Kelly. 2000. *Expanding the Mandate: Should the Border Environment Cooperation Commission and the North American Development Bank Go beyond Water, Wastewater, and Solid Waste Management Projects and How Do They Get There?* Austin: Texas Center for Policy Studies, July.

Reynolds, Steve. 1968. "The Rio Grande Compact." In Clark S. Knowlton, ed., *International Water Law along the Mexican-American Border*. El Paso: Contribution no. 11 of the Committee on Desert and Arid Zones Research, Southwestern and Rocky Mountain Division, A.A.A.S, pp. 48–62.

Secretaria de Relaciones Exteriores. 1947. *El Tratado de Aguas Internacionales*. Mexico, D.F.: SRE, Oficina de Límites y Aguas Internacionales.

Secretaria de Relaciones Exteriores. 1975. *La Salinidad del Río Colorado: Una Diferencia Internacional*. Mexico, D.F.: SRE, Collecion del Archivo Historico Diplomatico Mexicano, Serie Documental, Num. 13.

Snape, William J., Rene X. Acuna, and Federico Prieto Gaxiola. 2005. Petition and Notice Letter on Significant Legal Deficiencies in Present Proposal to Build and Line an Alternative All-American Canal in Imperial County, California. Washington, DC: Petition Submitted to Secretary Gale Norton, et al., U.S. Department of Interior, May 17.

Sohn, Louis B. 1970. Confidential Memorandum to Mr. Aldrich, Office of Legal Affairs, U.S. Department of State. Subject: Colorado River Salinity Problem: Reference to the International Court of Justice. NACP, Record Group 59, Mex-U.S., POL 33–1.

Spalding, Mark. 2000. "Addressing Border Environmental Problems Now and in the Future: Border XXI and Related Efforts." In Paul Ganster, ed., *The U.S.-Mexican Border Environment: A Road Map to a Sustainable 2020*. San Diego: Southwest Consortium on Environmental Research and Policy, SCERP Monograph no. 1, University of San Diego Press, pp. 105–138.

Timm, Charles A. 1941. *The International Boundary Commission, the United States and Mexico*. Austin: University of Texas Press.

Treaty between the United States of America and Mexico. 1944. Utilization of the waters of the Colorado and Tijuana Rivers and of the Rio Grande. 1944. 59 Stat. 1219, TS 994, February 3.

Treaty of Guadalupe Hidalgo. 1848. Treaty of Peace, Friendship, Limits, and Settlement between the United States of America and the Mexican Republic. [Treaty of Guadalupe Hidalgo]. February 2, 1848, 9 Stat. 922 (1851), T.S. no. 207.

Utton, Albert E. 1991a. "International Waters: International Streams and Lakes; Canadian International Waters; Mexican International Waters." In Robert B.

Beck, ed., *Waters and Water Rights*. Charlottesville, VA: The Mitchie Company, pp. 1–128.

Utton, Albert E. 1991b. "The Transfer of Water from an International Border Region: A Tale of Six Cities and the All-American Canal." *North Carolina Journal of International Law and Commercial Regulation* 16, no. 1 (Fall): 477–495.

Vaux, Henry. 2000. Plenary remarks at conference ". . . to the Sea of Cortés, Nature, Water, Culture, and Livelihood in the Lower Colorado River Basin and Delta." Riverside, CA, September 29.

Waller, Tom. 1992. "Southern California Water Politics and U.S.-Mexican Relations: Lining the All-American Canal." *Journal of Borderlands Studies* 7 (Fall): 1–32.

Ward, Evan. 2003. *Border Oasis*. Tucson: University of Arizona Press.

Zamora, Steve. 1971. Memorandum to Mark B. Feldman, American Republics Affairs: Legal Interpretation of the Term "From Any and All Sources" in the 1944 Water Treaty with Mexico. NACP, Record Group 59, Boundary and Water Disputes, United States and Mexico, POL 33–1, Mex-U.S. 10–1, August 27.

6

Developing a Plentiful Resource: Transboundary Rivers in the Pacific Northwest

Paul W. Hirt

Two decades ago, historian Martin Melosi wrote a history of energy in the United States titled *Coping with Abundance.* In it, he argued that the material fact of energy resource abundance in United States history (lots of wood, coal, oil, natural gas, and hydropower) profoundly shaped energy policy, energy systems, and public attitudes and behaviors toward energy production and consumption.[1] As Melosi (1985) argued, energy abundance supported rapid economic growth and industrial power but also spawned a great deal of waste and inefficiency along with a host of environmental and social problems poorly addressed by the laissez-faire politics that thrive in the presence of resource abundance.

The same observation might be made about water in the Pacific Northwest where abundant rainfall, deep snowpack, and prodigious river flow has created both a condition and a perception of plentitude for most of the history of the region. The widespread sense that the resources of the Pacific Northwest were inexhaustible led to grandiose dreams of an empire of wealth and liberty built on that natural abundance. While an empire did, in fact, arise, it was plagued with profligacy, pollution, and social inequality. The sense of abundance, however, convinced people that there was little need for government intervention or regulation, which made it exceedingly difficult to regulate exploitive and wasteful practices, mitigate externalities, avoid boom-and-bust economics, and address the social injustices that lead to political unrest.

The story I will tell here focuses on two major Northwest rivers—the Columbia and the Fraser—and the changing social, economic, and political landscape they have flowed through during the past 150 years. Throughout history, people have turned to the Columbia and the Fraser rivers primarily for four things: the water itself (for transportation, irrigation, waste disposal, etc.), the fish and wildlife that live in the water (salmon, beaver, otter, etc.), the kinetic energy embedded in falling water (water power), and the wealth that can be generated by controlling and allocating the water in those rivers. The complex and often competitive relationships that have developed over these four material resources have been further complicated by the many changing "boundaries" that people have created over time to divide the waters and to divide themselves. Political jurisdictions, social class, and racial identity have ineluctably structured the social landscape of the region; yet, the waters of the Columbia and Fraser remain transboundary phenomena in the physical landscape of the Northwest, continually eliciting cooperation or conflict among those divided from each other in the social landscape and in their use of the rivers.

The fates of these two rivers, and the international and intranational agreements that now structure their management and use, are embedded in historical context. Present conditions and ongoing equity concerns cannot be understood or resolved without understanding the evolution of river use and development over time. Because it would be impossible to cover all aspects of river use history in this chapter, I will focus on the key sources of conflict on these two river systems: the long-term tension between maintaining fisheries and developing an efficient hydropower system.

Physical context plays a significant role in shaping events and relationships. In contrast to arid regions where negotiations focus on who gets to divert water *out of the river* and for what purposes (see chapter 5 on the United States-Mexico water politics),social negotiations over the Columbia and Fraser rivers have focused more on what is done with the water and the fish *in the river*. Moreover, water *quality* issues are more significant in the Southwest where lower flows, irrigation runoff, sedimentation, and nonpoint pollution sources degrade water quality more extensively than in the water-abundant and well-vegetated Northwest.

Relationships between the United States and its two border nations are mirror opposites in the North and the South. In the North, with a few exceptions, the United States is the downstream nation trying to optimize its use of waters flowing from Canada; while in the South, Mexico is the downstream riparian trying to protect its claims to water flowing from the United States. The asymmetry in these relationships between the United States and its neighboring nations is much greater in the South than in the North. Canada has a stronger bargaining position than Mexico on water resources because it is the upstream riparian, it is a wealthier nation, and it possesses an abundance of water, which the United States covets.

In the United States-Mexico border region, as Stephen Mumme shows in chapter 5, governments and economic elites responded to water scarcity by aggressively capturing available resources, exercising tight controls, and adopting exclusionary policies regarding the rights and access of others. Rather than cooperation, intransigent self-interest dominated negotiations among competing water users. Stronger parties in the conflicts (e.g., the Imperial Irrigation District) secured early positions of advantage, held on to them tenaciously, and consistently denied any obligations to weaker parties (e.g., farmers in the Mexicali valley south of the international border). The resultant inequities and injustices led to decades of court battles, political gridlock, and lingering resentments.[2]

In the United States-Canada borderland, water and fish abundance supported a different set of relationships. Resource abundance created a culture of optimism and sustained a laissez-faire political economy. On the Columbia River, it also promoted more cooperation between nations and states than in the United States-Mexico borderland, at least insofar as developing an integrated hydropower system and negotiating water rights. There was conflict over the Columbia to be sure, but not like the conflict over the Colorado River in the Southwest. Because fishers and dam-builders in the well-watered Northwest do not take water out of the river for their exclusive use, decision-makers believed (perhaps naively) that they could harmonize these competing interests in the shared resource. And, because there were so many rivers and so much salmon, at least at first, all the parties believed that even if the two uses were incompatible, there was plenty of opportunity for both industries to thrive. By way of contrast, agriculturalists, cities, and states in the desert Southwest feared scarcity and engaged in cutthroat competition over water from the beginning. The battles in this region were not a matter of harmonizing a shared interest in abundant flowing rivers. Instead, it was a battle of the highest stakes over who got to capture, divert, and keep how much of the river for themselves.

Over time, of course, what was once seen in the Northwest as inexhaustible became the object of fierce conflict amid a rising sense of scarcity. Salmon were the first to shift from abundance to scarcity. Starting in the 1890s, salmon runs went into a steep decline. This led to a dramatic shift in fishery politics and growing conflicts between different classes of fishers as well as between fishing interests and the newly developing hydropower industry. This spawned the same kinds of relationships over fish as had occurred in relationships over water in the Southwest: competitors aggressively capturing as much as possible, contentious battles erupting over rights and access, exclusionary practices, inequities, and denials of any obligation to share the remnants with minority interests.

Although salmon were considered an increasingly scarce resource in the early twentieth century, water was not; so the politics of abundance continued to dominate river development affairs in the Northwest until the 1960s. An attitude of "we can have it all" optimism imbued both the private sector and government agencies as river development accelerated in the United States between the 1930s and 1950s. Cooperative river basin planning also thrived in these years, exemplified by the Bonneville Power Administration's (BPA) coordination of the regional electricity grid and a historic joint development treaty for the Columbia River signed by the United States and Canada in 1961. Ominously, however, the impact on fisheries from the hydropower system on the Columbia River accelerated in these decades, despite technological efforts to make the two compatible. At this point, the United States and Canada followed two different paths in attempting to resolve this conflict over efficiency and equity in the allocation and use of the Fraser and Columbia rivers. The reasons for their divergence and the different outcomes that resulted are the subject of this chapter.

Since physical context plays such an important role in shaping historical events, let us start by looking at the rivers themselves. The Columbia and the Fraser rivers constitute two vast and complicated transboundary watersheds that flow through two countries, one Canadian province, five northwestern states, and traditional territories of at least nine First Nations of Canada and twenty Indian tribes of the United States.[3] Besides this international geography, the rivers have other transboundary properties. Hydropower generated from the Columbia River flows in both directions across the international border, especially since the 1960s when the electrical systems of the United States and Canada were interconnected into a single massive electricity transmission grid that extends from British Columbia south to California and Arizona. Snowmelt in Canada fuels air conditioners in the desert Southwest.

There has also been a significant cross-border flow of capital associated with the flow of water and electrons, although contrary to the north-to-south direction of water flow, capital has mostly gone from

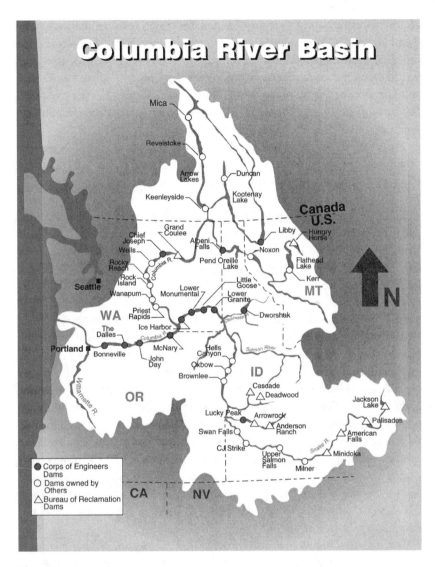

Figure 6.1
The Columbia Watershed. *Credit:* U.S. Army Corps of Engineers, Portland District

south to north to compensate Canada for river development activities that benefit the United States (more on this below). Finally, the Columbia and Fraser rivers provided spawning grounds for what was formerly the world's greatest anadromous fishery including all five species of Pacific salmon. Because these two rivers are both transboundary water systems, the salmon that depend on them become by association transboundary water resources too.

The Fraser River is not geographically an "international" river like the Columbia because it runs from its glacial headwaters in the Canadian Rockies and Coast Ranges to salt water entirely within the province of British Columbia. Nevertheless, because salmon are anadromous—they migrate between fresh and salt water over their lifetime—and because the route they take on their migrations goes through international waters in Puget Sound and the Strait of Juan de Fuca, which are accessible to fishers of both nations, the Fraser River sockeye run is indeed a transboundary resource.

The Columbia and Fraser rivers are truly remarkable in size and character—rivers of plenty indeed. According to the United States Army Corps of Engineers, "the Columbia River pours more water into the Pacific Ocean than any other river in North or South America": just under 200 million acre-feet of water every year (USGS 2002). Annually, it carries the second largest flow of any river in the U.S. (the Mississippi River is the largest). The Columbia flows more than 1,200 miles from its headwaters at Columbia Lake in the Canadian Rockies of British Columbia to where the river meets the Pacific Ocean on the Oregon-Washington border, draining 258,000 square miles. In contrast, the Colorado river drains a similar-sized watershed but yields an annual flow of less than 13 million acre-feet—only 6.5 percent of the amount of water in the Columbia. Canada contains 15 percent of the Columbia River watershed but contributes 25 percent of the river's flow.

This tremendous volume of water falling many thousands of vertical feet over its course makes the Columbia River the greatest hydroelectric power-generating river in the United States. In fact, the U.S. Bonneville Power Administration (BPA) claims it is "the largest hydroelectric system in the world" (Bonneville Power Administration, Multi-purpose dams n.d.). No new dams have been built in the American half of the Columbia watershed since the late 1970s, yet hydropower still provides 60 percent of the Northwest's electricity (Bonneville Power Administration, Power Generation n.d.).

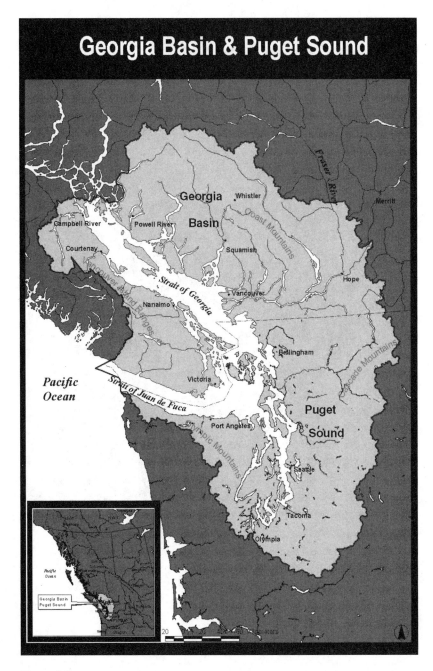

Figure 6.2
Strait of Juan de Fuca, Georgia Basin, and Puget Sound. Courtesy of Environment Canada.

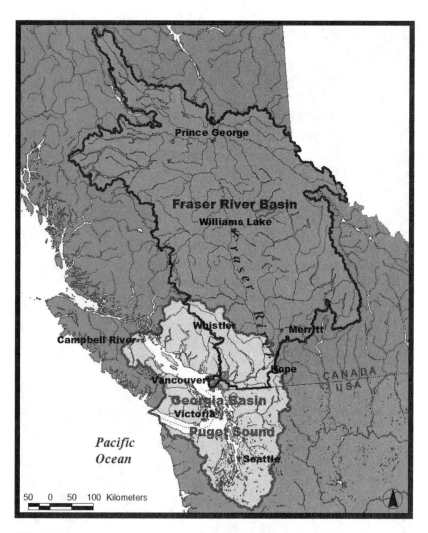

Figure 6.3
The Fraser Watershed. Courtesy of Environment Canada.

All this water flowing into the Pacific made the Columbia and other Pacific-draining rivers, like the Fraser, prodigious salmon streams. One hundred fifty years ago, before the salmon canning industry and the construction of large dams on the Columbia, an estimated 12–15 million adult salmon returned every year to the Columbia River and its tributaries to spawn. Now the number is closer to one million, with most of those hatchery-reared fish. The wild runs in most of the basin are federally listed as threatened or endangered. The government of the United States has spent billions of dollars in the past two decades trying to recover them from the brink of extinction—with little success.[4] The Northwest Power Planning Council estimates that about 80 percent of salmon mortality is traceable to the hydropower system. Consequently, electricity revenues subsidize the efforts to mitigate the impact of the hydropower system on fish and wildlife (Northwest Power Planning Council 1986; Blumm 1981).

The Fraser River in Canada is a fascinating contrast. It is the largest river in British Columbia. In fact, it has 50 percent more water in it than the Canadian reach of the Columbia River. Unlike the Columbia, the Fraser River is entirely *un-dammed* on its main stem. As a result, the Fraser is now the world's most productive salmon stream due to the fact that its main competitor for that title, the Columbia, has lost more than 90 percent of its historic fish runs. At 1,370 kilometers in length, the Fraser is also one of the longest undammed rivers in the world (Evenden 2004, 2–6).

How people understand these rivers, utilize them, and negotiate cooperation and conflict over them are determined significantly by culture, economic realities, political systems, class, race, and historical memory[5]—and some people in the region have long memories, especially those whose ancestors have lived here for many, many generations. Negotiating efficiency and equity in the allocation of the rivers and their resources is strongly influenced by memory and history. For example, resolving the question of who has the right to fish for scarce salmon on the Columbia River and how much they can take is strongly determined by several treaties between the United States government and sovereign Indian tribes negotiated in 1854 and 1855 (the Isaac Stevens Treaties). Two important United States District Court decisions in 1974 and 1979 (the Boldt Decisions) interpreted these treaties in ways that radically reordered fishing rights and the politics of salmon in the Northwest.

Likewise, determining how much federal dam managers should forgo efficiency in hydropower generation and how much electric utility ratepayers should pay to assist in the recovery of endangered salmon is determined largely by a 1980 federal law that based compensation on how much those river users are responsible for the decline in salmon runs over the past 150 years (more on this below). Historical estimates of how many salmon spawned in the Columbia and its tributaries prior to the arrival of the salmon canning industry and the construction of dams partly determines compensatory payments by electricity consumers today. History matters. Therefore, let us revisit past conditions and events that have shaped current events and that provide important context for what still concerns us today.

When European traders and explorers arrived in the Pacific Northwest in the late 1700s, the indigenous population along the coast was the densest of any region in North America north of Mexico. The societies inhabiting this relatively fecund land lived in densely settled villages, had a well-developed hierarchical social order, excelled in arts and handcrafts, organized their lives according to religious principles, engaged in trade, accumulated wealth, and made war on their neighbors. In these fundamental ways they were not much different than the Europeans who entered the Northwest in search of strategic alliances, trade, wealth, and religious converts.

The first set of sustained relationships that developed between indigenous peoples and Europeans revolved around commerce. Each side had products desired by the other and a relatively equitable exchange ensued in many locations for many decades, built mainly around the fur trade. Successful trade depended on cooperation, and cooperation was enhanced by the fact that neither side was in a position to completely dominate the other. During the fur trade era, a new, cooperative, mixed-race social order evolved—something Richard White refers to as aptly as the "Middle Ground" (White 1991). From the 1780s through the 1830s, the fur trade brought dramatic changes and challenges to the social order of the Native peoples. Yet Native Americans still outnumbered Europeans, retained sovereignty over their lands and lives, engaged in the trade largely by choice, and retained the ability to support themselves on local resources as they had done traditionally from time immemorial.[6]

By the 1840s, however, this "Middle Ground" had dissolved. Devastating population declines among Indians due to the spread of European diseases decimated their efforts to resist the expanding European empires and destabilized their social organization. Native peoples experienced increasing economic vulnerability, competition over access to traditional lands and resources, and inequitable trade relations. The era of cooperation, intermarriage, and mutual commercial benefits was replaced by economic dependency, EuroAmerican political and racial hegemony, conflict, dispossession, and isolation of Native peoples from the mainstream of the new evolving political, social, and economic order in the region.

Legacies of this era of conquest have come back to haunt present negotiations over water resources, as the inequities of the nineteenth century are only slowly being redressed.[7] In the American Northwest, a series of treaties were forced on the Coastal and Plateau tribes in 1854 and 1855 by Isaac Stevens, the first territorial governor of Washington. Stevens insisted the Indians cede their claims to most of the territory so that those lands could be legally opened to acquisition and settlement by non-Indians. Tribal leaders opposed to signing the land cessions were summarily dismissed by Stevens from the treaty negotiations and replaced by more amenable leaders appointed by Stevens. Resistance to the terms of the treaties dictated by Stevens were met with threats of military reprisal.[8] The main "enticements" for Indians to sign the treaties were a promise of peace (not honored), offers of permanent "reservations" for certain tribes on which they would exercise some degree of self-determination (the permanence of the reservation boundaries was not honored nor was there much self-determination), annual payments for a fixed period of time as compensation for ceding tens of millions of acres to the United States government, and, perhaps most importantly, "the right of taking fish at all usual and accustomed grounds and stations. . . ." (also rarely honored during most of the nineteenth and twentieth centuries).[9]

This right to take fish at usual stations was critical. The fish being referred to were salmon, which typically made up 30–80 percent of the diet of Northwest Natives along the coast and in the interior river basins where salmon spawned. Coastal Indians built their villages near fishing grounds and jealously guarded their best fishing sites. The more mobile interior plateau Indians returned every spring and summer to traditional fishing sites, which they often shared with other bands. During the few months of salmon runs, they collected and dried all they needed for the remainder of the year.[10] The truly spectacular fishing grounds like Celilo Falls and Kettle Falls on the main stem of the Columbia River, where millions of salmon gathered in pools at the base of the falls seeking opportunities to leap the basalt barriers on their upstream migrations, were sites of huge annual trading fairs and summer social gatherings for Indians from the entire region.

Trade goods from the coast to the Great Lakes have been unearthed at archaeological sites at Celilo Falls. Salmon as food, as a trade item, and

Figure 6.4
Dipnet Fishing at Celilo Falls, Columbia River. *Credit:* Washington State Historical Society, Tacoma. Photo by Virna Haffer. Negative number 1974.35.2261.39.

as a religious icon were absolutely central to Northwest Natives' lives, economies, and cultures.[11]

Stevens, of course, knew that the tribes would never agree to giving up their claims to 64 million acres of the Pacific Northwest without some guarantee that they would continue to have access to salmon fishing grounds and rights to all their other traditional subsistence activities: to gather shellfish, hunt deer and elk, and collect roots and berries at usual and accustomed places. All his treaties contained language guaranteeing this right to access key subsistence resources on and off the reservations. It was not an exclusive right, but rather "in common with citizens of the United States." This guarantee, however, was regularly ignored or denied by state and federal governments and non-Indian individuals from the time the Indians moved onto their reservations until federal court decisions of the 1970s resurrected the treaties from the dustbin of history and assigned to Northwest Indian tribes the rights to fully half the annual salmon harvest. By then, however, Grand Coulee Dam had drowned Kettle Falls, The Dalles Dam had drowned Celilo Falls, and the Columbia River fishery had collapsed (more on this below).

In pursuit of the most efficient means to open up Northwest lands to non-Indians, equity considerations played a small but important role in treaty negotiations but were not honored afterward. Because the balance of power had by the 1850s tilted substantially toward the government of the United States, the Stevens treaties were not a negotiation among equals; they were a thinly veneered instrument of conquest and dispossession. Arguing for the importance of equity, we can see that the unilateral imposition of the treaties to the disadvantage of the Natives and the failure of the United States to subsequently honor most of the clauses that benefited Native peoples led to a century and a half of Indian poverty, dependence, resentment, and social conflict, which has become a moral embarrassment, a financial burden, and now a legal quagmire, as well as an epic human tragedy for those adversely affected for so many generations.

Nineteenth-century whites mostly hoped that the "Indian problem" would simply go away, and these inequitable treaties were a means to diminish the Indian presence and influence on the landscape. Yet, Indians did not go away, and the problems associated with conquest and dispossession merely festered (Limerick 1987). Recently, the thin veneer of legality embodied in the instrument of dispossession—the ignored promises in the Stevens treaties—have been resurrected in a bid for justice that offers at least the possibility of righting the wrongs of the past and fostering a more desirable, sustainable, and mutually beneficial relationship between Natives and newcomers in the Northwest. But this redress of inequities did not come until well into the twentieth century, and then only in halting incremental steps. In the meantime, the inequities piled on for Native peoples, while whites in the Northwest turned their attentions to their own set of water resource concerns following the confinement of Indians to reservations.[12]

Native peoples were not the only ones fishing for salmon in the nineteenth-century Northwest. Virtually everyone—trappers, traders, merchants, farmers, miners, Irish, Chinese, Hawaiians, men and women, young and old—fished for or ate salmon at least part of the year. Selling fresh or dried salmon supplemented many people's incomes. Even the Hudson's Bay Company marketed dried salmon as one of its trade staples. Salmon were so important to the formal and informal economy of the Northwest that the Oregon Territorial Constitution of 1848 actually included a prohibition against anyone creating dams or obstructions on salmon-bearing streams.[13] This clause in the territorial constitution was an early foreshadow of a conflict over competing river uses that would animate Northwest politics for the next 150 years. Unfortunately, its provisions were ignored the way treaty promises to Indians were ignored. No mention was made of fish at all in the 1859 Oregon State constitution.

Government efforts to sustain this important resource later in the nineteenth century through regulations and fish hatchery construction also did not protect the fish from their most voracious predator: commercial fishermen working for the salmon canning industry. Shortly after the Civil War ended, the industrial and market revolution that had already transformed the eastern half of North America arrived in the Northwest. The first salmon canning operation in the Northwest was established in 1866 by two Sacramento canners looking for more abundant runs after the canning industry and hydraulic mining had devastated the salmon runs on the Sacramento River in central California. They set up operations near what is now Astoria, Oregon at the mouth of the Columbia River.

Within two decades, several hundred canneries had sprouted up along the Columbia River and its tributaries. The efficiency of these operations in capturing, canning, and marketing the annual salmon runs was enhanced by new technologies such as fish wheels and mechanical slicers, and by faster and cheaper transportation (steamships and railroads) bringing canned salmon to the booming urban populations of the United States, Canada, and Europe. Mass production was the new order of the day, and in this Gilded Age climate of frontier boosterism and economic libertarianism, the fishermen working for the canning industry quickly decimated the salmon runs on the Columbia. Lack of self-restraint or effective external controls over this common pool resource doomed any dream of a sustainable fishery (Goble and Hirt 1999, 229–63).

The state of Oregon offers a representative example. It passed its first fish and game laws in 1872, including a requirement that all dams must have fish passage facilities, but the legislature appropriated no money and created no agency to enforce those laws for the next fifteen years. In 1887, the legislature finally established a Board of Fish Commissioners and gave it some meager funds to operate a couple of fish hatcheries. In 1898, after the Columbia River salmon harvest had already begun its dramatic decline, the Oregon legislature passed a more comprehensive salmon protection law, but typically it was too little too late.[14] The measures authorized were ineffective, poorly enforced, or concertedly resisted by those inconvenienced by the salmon protection regulations. By the late 1890s many canneries along the lower Columbia had shut down, and businesses were relocating farther north to Puget Sound and British Columbia in search of unexploited salmon runs.[15]

After 1915, despite the vigorous efforts of the Washington State Fish Commissioner, salmon populations in the Puget Sound Basin were in serious decline due to overfishing, pollution from lumber and pulp mills that lined the harbors, destructive logging in the watersheds that silted up streams and spawning beds, and dams that blocked salmon migration. Fishermen and canners edged farther north once again to better fishing sites in Canada.

British Columbia's experience was similar to the United States at this time, though lagging a decade or so behind in the boom and bust sequence. The first cannery on the Fraser River was established in 1871 and after about two decades observers noticed a slow but steady decline in that fishery. The bust came rather suddenly after 1913 as a result of a monumental series of rockslides in the steep, narrow stretch of the Fraser River canyon called Hell's Gate. These rockslides were caused by hurried construction of a poorly engineered railway through the canyon. The slides almost completely blocked salmon migrations upstream for the next eight years and despite many efforts to clear a passage through the rock debris, the sockeye run did not experience any significant recovery until the late 1940s when major restoration work was completed. Fishermen on the lower river continued to capture salmon, but the numbers dropped precipitously since the upper river's spawning areas were no longer accessible for reproduction (Evenden 2004, 19–52).

This migratory boom and bust pattern in the salmon industry is a classic result of the tragedy of the commons. Oregon, Washington, British Columbia, and the federal governments of both nations had all created laws and agencies in the late nineteenth century dedicated to sustaining the fisheries, regulating those who used it, and protecting their interests against conflicting uses of the rivers, but to little avail.[16]

While overfishing constituted the main cause of the boom-and-bust cycle in the early twentieth century, this was not the only problem facing the salmon fishery. As the economy of the area developed, especially after the arrival of the transcontinental railroads in the 1880s and 1890s, many other resource extraction activities accelerated in the Northwest, and many of those new industries also made use of the Northwest's rivers.[17] The timber industry built narrow-gauge railways up river valleys and logged everything they could reach, leading to extensive soil erosion during the nine-month rainy season. Loggers often used streams and rivers to transport logs, with devastating impact on the riparian areas.

Mining often had devastating effects on rivers and fish, too. Crushed ore and tailings piles eroded into streams during rains and filled riverbeds with sand and gravel. Toxic metals from mines leached into the rivers, poisoning aquatic life and the people and livestock drinking from the stream. Chemicals for processing minerals from ore were simply dumped after they were used or no longer needed. They made their way into watercourses, too, as did the pollutants from the human populations in the mining boomtowns that provided the labor and services for those mines.[18] Hydraulic mining especially destroyed riparian vegetation and sent sand and gravel downstream for miles clogging riverbeds with sediment.

Mining companies also damaged fisheries in their efforts to secure water and power for their operations. In British Columbia, for example, a large mining company built a dam on the Quesnel River in 1898, a significant salmon-bearing tributary of the Fraser, which blocked one-quarter of the river's spawning grounds and significantly reduced the sockeye run. The mining company built the dam despite the protests of the salmon canning industry and other fishery advocates. Federal and provincial government agencies did nothing to demand effective fish ladders or other mitigations or compensations (Evenden 2004, 69–76).

A basic element of equity, in the case of common pool resources such as rivers and the fish in them, is that one person's use of those resources should not preclude or damage another person's use or enjoyment of them, without formal assent and compensation for losses. Unfortunately, the Oregon territorial constitution and subsequent state and provincial laws designed to sustain salmon runs did not in fact prevent the building of obstructions on rivers that harmed or blocked salmon migrations, and those who used or depended on the fish usually had no effective recourse and received little or no compensation, at least not until the second half of the twentieth century.

Under earlier American and Canadian water law, of course, harmed parties did have some recourse. A new water-law system evolved in the late nineteenth century, however, to facilitate privatization of water, particularly in the arid western regions of North America. The new water rights doctrine served very effectively to advance the interests of industrial capitalism (Robbins 1997, 110–204). This crucial water policy innovation of the nineteenth century, which remains at the foundation of water use and management in the arid West today, was the replacement of common law riparian rights with the Doctrine of Prior Appropriation.[19] In the riparian system, water was used but not owned; it was a community resource that served a variety of purposes. No one's use could diminish another's use without the harmed party having the opportunity for redress or compensation—at least in principle.

In contrast, prior appropriation law says the first person to divert a stream for beneficial use has priority claims to that water to the exclusion of others: "first in time, first in right." Both Canada and the United States adopted this new water rights regime in their western regions, although Canada adopted it less enthusiastically than the United States, and government authorities retained more regulatory oversight in the development of water resources than in the United States.[20] Such a system helped industrialists, farmers, and ranchers in the newly developing regions secure their rights to water without having to worry about sharing it, cleaning it, or even returning it to the riverbed. In fact, in the doctrine of prior appropriation, diverting water is an explicit condition for establishing a claim to it. Irrigators with this new kind of water right could dam and divert river water with relative impunity; in some cases, irrigators completely dewatered streams leaving those downstream high-and-dry and ruining fish and wildlife habitat.[21]

While prior appropriation policy certainly stimulated private enterprise, it turned out to be very ill-suited to achieving such goals as efficiency, equity, community participation in decision-making, and sustainability. Moreover, though it gave priority rights holders a clear legal advantage over others, it only exacerbated the problems associated with conflicting uses of rivers and environmental degradation. A large segment of the public had little or no ability to establish or protect water rights claims either for themselves or on behalf of the public interest. Those dependent on fishing had no legal standing in water rights cases since they did not "divert water for beneficial use." Their livelihoods and that of the fish depended on clean water continuing to flow in unobstructed rivers, yet they were unable to claim a right to clean flowing rivers under western water law. Later settlers faced a stranglehold on water resources held by earlier arrivals. The poor and marginalized, even those who were in fact there *first*, were disadvantaged in claiming water, diverting water for their use, and protecting their interests in court.

Lingering ideological resistance to government regulation and restraint, especially in the United States, hampered the efforts of government fishery agencies to compel those who damaged the water commons to alter their activities or desist. The decline in salmon runs made matters worse, since the canning industry lost political clout in the same proportion that it lost economic position in the state and provincial economies. As a consequence, fish interests adopted three main strategies: (1) fishermen themselves tried to curtail access and competition in the fishery; (2) regulators placed their hopes in artificial enhancement of the salmon runs through hatchery production; and (3) both organized to keep dams and other development out of the most productive salmon-spawning rivers.

Regarding competition, private associations of economic interests often organized themselves to restrict others from gaining access to salmon. The Columbia River Fisherman's Protective Union, for example, was incorporated in 1884 to control who fished the lower Columbia River. Mainly an association of Scandinavians and Finns who owned fishing boats on the lower Columbia, the Union divided up the fishing grounds among themselves and kept competitors out, especially Asians (Martin 1994; Witt n.d.). Similarly, a Fraser River Fishermen's Protective and Benevolent Association was formed in 1893 to exercise control over who fished the lower Fraser. Japanese fishers were barred from joining the association. A plank of the association's charter explicitly stated that one of its goals was to exclude the Japanese fishing fleet from the Fraser fishery. In response, the Japanese formed their own rival organization, Gyosha Dantai, but were forced to fish less-productive areas in open water not easily controlled by others (Isabella 1999).

In addition to trying to exclude Asians from fishing (though they were generally welcomed as laborers in the canneries), whites also dismissed Indian fishing as a mere "subsistence" activity, not worthy of protection in times of scarcity. For example, after the landslides in Hell's Gate on the Fraser River in 1911–1914, the cannery interests demanded that the provincial government curtail the Indian fishery (but not their own) until the slide could be cleared and the runs recovered. The government complied with this request, but the Indians refused to submit to its order not to fish. They correctly pointed out that their catch was miniscule in comparison to the harvest taken by the canners, that the order to desist unfairly targeted them when they had done nothing to cause the slide and its decimation of the sockeye run, and that in any case the desist order violated their long-established rights to fish. First Nations groups organized successful protests and civil disobedience and made some headway in advancing their claims for equity during the 1910s (Evenden 2004, 36–43).

Efforts of Indian tribes in the United States to secure greater justice and rights met with less success at this time. In fact, the Indian fishery in the United States was regularly restricted by state fish authorities, increasingly so as the salmon runs diminished, in favor of non-Indian commercial and sports fishing interests throughout most of the twentieth century. The turning point came with David Sohappy's act of civil disobedience in the late 1960s—fishing for salmon in defiance of state regulations and in defense of treaty rights—which led to the case of *Sohappy v Smith* (302 F. Supp. 899, D. Oregon 1969) and the aforementioned Boldt Decisions in the United States District Court during the 1970s.[22]

The conflicts over access to the salmon fishery had a strong international dimension, especially regarding the Fraser River. Canadians were constantly pressing for an international agreement that would limit capture of the Fraser sockeye runs by fishers from Puget Sound as the adults swam through United States and international waters on their return to spawn in the Fraser River. Canadian studies from the 1920s estimated that Puget Sound–based fishers took about two-thirds of the Fraser sockeye run. Since Canadians invested all the expense and effort to sustain that fishery through habitat improvements and hatcheries, it seemed unfair for American fishers to take the bulk of the harvest. Three bilateral attempts to coordinate United States-Canadian salmon management failed consecutively in 1892, 1908, and 1919. A fourth attempt in the 1930s finally succeeded.

After seven years of haggling, the two nations signed a convention in 1937 establishing the International Pacific Salmon Fisheries Commission (IPSFC). This organization's efforts to address equity issues could easily be the subject of an entire study. In short, the IPSFC was at first hamstrung by a requirement that it study fishery problems for seven more years before taking any actions or issuing any regulations, so its active management of the Fraser salmon runs did not really begin until 1943. Its efforts focused on improving salmon passage on the Fraser, determining the timing of sockeye runs in order to establish limited closures during the run to allow some fish to reach their spawning grounds, and restraining United States fishers from taking the majority of the catch (Evenden 2004, 84–91). But antiregulatory protests by United States-based fishers and ongoing disputes over how to determine which fish belonged to which country continued to hamper progress on international cooperation. In an attempt to effectively manage this shared transboundary resource and equitably divide the catch, the United States and Canada signed several additional salmon conventions since 1937, most recently in 1985 and 1999. Heated confrontations between Canadian and American fishers still recur.[23] It seems likely that as long as one side or the other perceives the division of the resource as unfair, controversy will persist.

In the face of declining salmon populations, due to the inability to restrain the fishing industry or control damaging activities in the rivers where salmon spawned, state, provincial, and federal fish authorities turned to a second strategy—the technological fix. Starting in the late nineteenth century, fish managers aggressively promoted salmon hatcheries and fish passage technologies at dams. By building fish propagation facilities that reared and released millions of young salmon yearly and requiring dams to have fish ladders, managers hoped to stave off the total collapse of the fishery. The first hatchery in Oregon was built by the United States Fish Commission in 1877 on the Clackamas River. Within one decade, hatcheries became an increasingly popular state-funded response to salmon decline. Washington State was fully committed to hatcheries by the 1890s. British Columbia imported hatchery technology and expertise from the United States at the turn of the century. Rules regarding fish passage at dams, however, were much more difficult to enforce, and fish-ladder technology was poorly understood and often ineffective, especially as the size of dams increased during the twentieth century.[24]

More important than the material output of the hatcheries, fish propagation offered politicians and regulators hope that the fishing industry could be sustained without recourse to strict regulations, government rationing, or curtailment of other river development activities. Hatcheries supported a conspiracy of optimism that served only to postpone the inevitable reconciliation of conflicts (Lichatowich 1999, 194). Salmon hatcheries are still used extensively today on the United States side of the international border and salmon populations there are still threatened with extinction. Ironically, after the 1920s the government of British Columbia mostly abandoned the use of hatcheries as a solution to fishery decline, yet today the Fraser River supports the largest sockeye run in the world. The reason for these differences is that British Columbia followed a separate path from the United States in pursuing the third strategy for sustaining salmon runs. It did not delude itself with technological optimism. Instead, it protected the social value of salmon by avoiding clearly incompatible developments in important salmon habitats.

The third strategy of fishery interests was to advocate for land and water protection designations—such as parks, wildlife refuges, fishing closures, and the like—and to oppose the developments that harmed salmon runs and natural salmon reproduction. The national governments and the states and provinces established an impressive collection of nature preserves and conservation initiatives designed to protect fisheries in the Northwest. Between 1890 and 1920 Congress and the president established about 50 million acres of national forests, as well as a half-dozen national parks and wildlife refuges in Washington, Oregon, Idaho, and Montana.

Canada acted similarly, designating numerous national and provincial parks in British Columbia during these decades. The provincial government established its first and still largest provincial park, Strathcona (on Vancouver Island) in 1911. Only a decade later, however, Strathcona was the subject of controversy over a proposed dam that the timber industry wanted to build in the park to provide power for a large mill, town, and related processing facilities. Conservationists were able to defeat that proposal in the 1920s. More often, however, the British Columbia government approved applications by the private sector to build dams, primarily for hydropower, during these first few decades of the twentieth century. Numerous small hydropower dams were built on tributaries of the Fraser River in these decades, and none of them had effective fish passage facilities. Consequently, each one decimated the salmon population in that stretch of the river and emboldened fish advocates to oppose further dams. During the 1920s, the Fish Commissioner for British Columbia, J.P. Babcock, announced that hatcheries and fish ladders were futile stopgap measures, and that dams and salmon were fundamentally incompatible. The only way to ensure a healthy salmon fishery was to maintain a natural river (Stadfeld 2004). Policymakers in the United States never took such a position, continuing to believe that it was possible to have an abundance of both dams and salmon.

Fishery interests in British Columbia faced formidable opposition from the mining, timber, and electric power industries, supported by the

British Columbia Water Branch, which had become especially active in promoting waterpower development. In fact, the Water Branch actively sought to weaken the political leverage of fishery interests and worked against the provincial fish authorities. This array of powerful opponents motivated British Columbia fishery interests to form a united front, with the cannery associations particularly active and influential. Still, they lost more battles than they won (Evenden 2004, 157–170; Stadfeld 2004, 73–84). The economic depression halted most dam proposals during the 1930s at a time when the main stem of the Fraser as well as the Canadian reach of the Columbia River remained free of dams. Fishery advocates in Canada had about 15 years of breathing room before the pressure to dam the Fraser resurfaced.

Conservationists in the United States ironically had a mixed agenda that included fish and wildlife protection as well as aggressive river development, with little apparent recognition of the conflict between the two. Besides establishing parks and refuges, the federal and state governments promoted irrigation, hydroelectric power, dredging of river channels, and other activities to maximize the utility and efficient development of water resources. With the electrical revolution, progressives were particularly interested in promoting hydropower as a means to "modernize" both industry and society.

As in British Columbia, fishery interests were a weak third-party in river management decisions. But unlike Canada, the fish advocates in the United States rarely blocked any dam proposals in the early decades of the twentieth century and they often failed to present a united front. As each new dam was built, the fishery declined more; yet sports fishermen bickered with canners, canners bickered with regulators, regulators focused on hatcheries and fish ladders, and everyone ignored or disenfranchised the tribes. Conventional wisdom in the United States assumed that the rivers would be developed. The only question was when and how. In Canada, developing every good dam site was not quite a foregone conclusion.

During the Great Depression, the United States and Canada took different paths. Despite the hard times, Canada continued to rely predominantly on the private sector for economic development, which meant that in the 1930s investment remained at a very low level. In contrast, the United States government, as part of Franklin Roosevelt's New Deal, invested billions of dollars in conservation works that employed laborers while developing rivers and other natural resources. In Canada, provincial authority remained as strong as, or stronger than, federal authority in the management of natural resources throughout the twentieth century. In the United States, federal authority grew continually stronger until by the 1930s it overshadowed the states in natural resource decision-making, especially in the West where the federal government owned much of the land base.

While Canadian river development languished in the 1930s due to a lack of will and investment capital in the private sector, the United States government poured large sums of money into federal development agencies like the Army Corps of Engineers, the Bureau of Reclamation, and the Bonneville Power Administration—all three engaging in massive river development schemes in the Northwest. This development boosted the economy and electrified the region but devastated the remaining fish runs and dispossessed tribes of their best remaining fishing sites. This made it nearly impossible for them to exercise their treaty rights to this resource that was so central to their commercial, subsistence, and religious lives. The decision to sacrifice one set of river values for another—sacrificing equity and minority rights for efficiency and majority benefits—occurred mainly because the harmed parties lacked sufficient political power to resist. The injustices would come back to haunt the government and river managers half a century later.

One of the first megaprojects undertaken by the United States government during the New Deal was the construction of Bonneville Dam on the lower Columbia River, only the second dam ever attempted on the main stem of the Columbia. Private investors had started construction of the first dam on the Columbia main stem, Rock Island Dam in the middle reach of the Columbia, just a few years earlier and completed it in 1933. Because it was not a very tall dam, Rock Island had fish ladders built into it. But they were poorly designed and this led, after the dam was completed, to an 88 percent drop in salmon taken by Colville Indians upstream at Kettle Falls.[25]

The Army Corps of Engineers started Bonneville Dam in 1933 and completed it in 1938. Congress created the Bonneville Power Administration (BPA) in 1937 as a temporary agency to market power from Bonneville Dam, but the BPA quickly grew to dominate and integrate the entire regional electrical generation and transmission grid for both public and private power (Tollefson 1987; Norwood 1981). Because of advocacy from fishery interests, Bonneville Dam was designed with better functioning fish ladders than Rock Island. While those ladders worked reasonably well to get adult salmon upstream, passage for the downstream migration of young salmon smolts going back to the sea remained very problematic. Ladders are designed to help salmon get upstream, not down, so downstream migrants pass over spillways or through turbines, and the mortality is extraordinarily high. Fewer young salmon surviving their journey to the ocean meant fewer adult salmon returning to spawn. Then (as today) this represented a serious obstacle to sustaining the salmon runs.[26]

Indians, once again, lost a great deal even though Bonneville's fish ladders worked to get some adult salmon upstream. The reservoir behind the dam drowned thirty-five Native fishing sites, including the third most important fishing site on the Columbia River—Cascade Falls. In compensation, the United States government promised to provide the affected tribes alternative fishing sites. But not only did the government fail to do so, it continued building dams that quickly and irreversibly decimated the entire salmon fishery of the Columbia and its tributaries (Ulrich 1999; Center for Columbia River History n.d.; Netboy 1980).

Bonneville Dam was only a piece of the development puzzle. Government planners resurrected an earlier report by the Army Corps of Engineers (USACE) proposing comprehensive "multi-purpose" river basin development in the main watersheds of the United States, including the Columbia (the 308 Report). Multipurpose at this time meant five specific things: navigation, flood control, hydropower, irrigation, and recreation. While recreation included hunting and fishing, engineers and policymakers considered salmon expendable (Netboy 1980).

Part of the comprehensive river basin development vision for the Columbia River included large storage dams in the dry interior plateau to hold some of the prodigious spring runoff for later release in the low-flow summer months. Some of these dams would also serve to divert water to irrigate up to a million acres of sagebrush desert in the rain shadow just east of the Cascade Mountains. Hydropower revenues from these dams would pay back the construction and maintenance costs of the projects and subsidize the pumping of river water from the reservoirs in the canyon up to the irrigable benchlands above it.

The Grand Coulee Dam was the lynchpin for this upstream storage and irrigation vision. The Bureau of Reclamation (BOR) launched this project shortly after the Army Corps of Engineers started Bonneville Dam. The Grand Coulee Dam was at the time the largest man-made structure on earth, set into the basalt walls of the Grand Coulee on the upper portion of the American reach of the Columbia River. Completed in 1941, the Grand Coulee Dam created reservoir slackwater more than one hundred river-miles to the Canadian border. Designed too tall for fish ladders, it was built with the full knowledge of what that meant to the salmon fishery. The Grand Coulee Dam permanently blocked all salmon runs to the entire upper watershed behind it—one-third of the river's spawning grounds and the entire reach in Canada. Moreover, the dam inundated Kettle Falls, the second most important Indian fishing site on the Columbia River, as well as most of the Colville reservation's farmland and several of its towns (Pitzer 1994). Compensation was, as usual, both meager and long-delayed. Protests against the dam from fishery interests on both sides of the border, including many tribes, conservationists, canners, and biologists, had no effect, even though the United States government had drawn up an alternative development scenario of several lower height dams that would produce the same amount of power but allow fish passage. This decision to build Grand Coulee Dam revealed an overwhelming government commitment to viewing water as a tool for building wealth in certain favored segments of the economy, rather than viewing water as a shared ecological resource providing broad social and environmental values.

Another monster dam was soon built below Grand Coulee. Finished in 1955, and backing water up all the way to Grand Coulee Dam, it is the second greatest producer of hydroelectricity in the United States (following Grand Coulee), and it, too, supplies federally subsidized irrigation water to farms in the Columbia Basin. Like Grand Coulee, it was a high dam constructed without fish ladders, which ended the salmon fishery in its stretch of the river. In one of the stunning ironies of American history, this dam was named after Nez Perce Chief Joseph.

The coup de grâce to the Native fishery in the Columbia Basin came in the mid-1950s with the completion of The Dalles Dam on the lower Columbia River above Bonneville. Like Bonneville, it was low enough to have fish ladders; but also like Bonneville it permanently buried critical Native fishing sites. In fact, it inundated the most important Native fishery on the river: Celilo Falls (Barber 2005). Some salmon made it up the Columbia past Bonneville and The Dalles, but there was no good place for Indians to catch the fish in the deep, placid reservoir. Worse yet for the long-term survival of salmon, fewer and fewer young smolts survived the treacherous passage downstream past these and other dams back to the sea.

Among the few remaining areas of the Columbia basin that still supported healthy and harvestable populations of salmon were the Snake River and Salmon River in eastern Washington and northern Idaho. An epic battle ensued in the 1950s and 1960s over proposed dams in those important tributaries of the Columbia, with a growingly influential environmental movement finally acquiring the political clout to play a significant role in decision-making. Aided by the conservationist Senator Frank Church of Idaho, the Salmon River was spared (Brooks 2006), but the Snake River was fully developed, including four dams on the lower Snake built between the 1950s and 1970s. Since the Salmon River is an upstream tributary of the Snake River, its salmon runs were decimated by the lower Snake River dams, even though the Salmon River itself is undammed and flows through one of the largest wilderness areas in the United States.

Both the Snake and the Salmon rivers and many of their tributaries flow through the Nez Perce Indian reservation and through the much larger traditional territory that the Nez Perce controlled before the United States government unilaterally shrank their reservation by 90 percent in 1863. In a hopeful sign of incremental justice and greater equity, Nez Perce Tribal fish managers are now key players in intergovernmental efforts to protect the remnant salmon populations in northern Idaho and have led the way in habitat restoration, innovated new small- scale hatcheries that mimic natural conditions, and lobbied effectively for salmon-friendly dam operations. They were also among the leaders of a surprisingly influential movement in the 1990s advocating removal of the four lower Snake River dams in order to recover salmon populations in the watershed (Landeen and Pinkham 1999). The Snake River dams remain in place, but a very serious and honest evaluation occurred regarding the broader costs and benefits of those dams, and their future remains uncertain.

In the cultural and political milieu of the United States from the 1930s through the 1950s, salmon were considered expendable—despite the fact that the United States had treaty obligations to Indians, spent millions of dollars trying to sustain the fishery, and passed numerous laws promoting fish and wildlife conservation. During the New Deal and for the two decades which followed, an economic calculus of maximum net benefits drove resource development decision-making. It was also a matter of national pride to be building what Woody Guthrie called, the "biggest thing on earth" (Grand Coulee). Economists and engineers determined that Grand Coulee, Chief Joseph, and other similar megadams would be the most *efficient* way to irrigate the dry Columbia Basin and produce cheap hydropower. They and many others argued that this was the best way to pursue the general social welfare, and if a "minority" were harmed, it was for the greater good.

This particular vision of economic efficiency and social welfare certainly served to provide an array of benefits to residents of the region, but those benefits were not equitably distributed. Indians, for example, lost much more than they gained (in land, fish, autonomy, security, and culture). When they got electricity or irrigation water from these dams it usually came later and less reliably than it did for whites, cities, and industry. The bias in the old river development calculus toward certain kinds of values (hydropower and irrigation for example) created a lopsided development that caused problems that must now be redressed. The failure to allocate resources equitably among the diverse population created resentments that haunt social relations today, losses that tie up the courts with compensation claims, and a social movement to undo some of the development that the government only recently completed (the lower Snake River dams).

Canadians watched all of this transpire with great interest. They saw that United States development of the Columbia resulted in abundant and cheap electricity, industrial development, and rapid economic and population growth. The lack of hydropower development in British Columbia in the 1930s, moreover, resulted in that province grudgingly having to import power from Washington State to meet its own electricity shortage in the 1940s. Canadians, always sensitive to inequitable or dependent relationships with its powerful southern neighbor, resented having to buy power from the United States when it had so much untapped hydropower potential of its own. Consequently, politicians and development boosters in British Columbia cast their eye toward the most abundant and readily available source of hydroelectricity near their urban centers: the Fraser River.

At the same time, other Canadians also watched the salmon runs on the Columbia make their final precipitous collapse following all the dam construction of the 1930s–1950s, just at the time the Fraser River runs were finally rebounding from the losses incurred after the landslides and overfishing earlier in the century. The lesson to fishery interests was crystal clear, underscoring Fish Commissioner Babcock's earlier position about the fundamental incompatibility of dams and salmon. The stage was set for a major battle over the future of the Fraser River, as river developers and their political allies were pitted against fishery interests and their allies in postwar British Columbia. The conflict, influenced greatly by interests and pressures from the United States, would be settled by 1958.

Starting in the 1930s and accelerating in the 1940s, the United States increased pressure on Canada to develop water storage reservoirs on the upper Columbia River. River and power managers in the United States wanted Canadian storage dams to hold back some of the Columbia's spring runoff and release it during times of low water flows in the late summer and fall to improve year-round hydropower system performance in the United States. The American government tried to push this agenda through the International Joint Commission (IJC), which was formed in 1909 as a forum for addressing United States-Canada border policy (Chacko 1968). The IJC had been virtually inactive in the Pacific Northwest up to that time, but in 1944 it was charged to investigate the potential for "joint development" of the Columbia River.

At the time, the government of British Columbia was unenthusiastic about developing the Columbia River. Policy makers and planners considered it too remote from the province's urban centers. Despite vigorous opposition from fishery interests, the provincial government instead encouraged several proposals for dams on the Fraser and its tributaries in the 1940s. Running parallel to this, Canadian federal government officials, under pressure from the United States government, lobbied British Columbia to cooperate in joint development planning for the Columbia River. Adding urgency to these pressures, a devastating flood occurred on the Columbia in 1948, the second biggest on record, underscoring the potential value of large storage dams which might be able to hold back future floodwaters and reduce downstream damage.

For the next decade and a half, the United States and Canada negotiated over how costs and benefits from joint development of the Columbia would be allocated. Negotiators sought to determine the value of the downstream benefits to the United States from Canadian storage of river water and to determine who would pay for and build these large storage dams. They also debated who would own, buy, and sell the extraordinary surplus of electricity generated at these dams. British Columbia did not need all that power. Moreover, it could not afford the cost of building the Columbia dams. It was more interested in developing the Fraser River, which offered no benefits to the United States. Consequently, much of the negotiations concentrated on what incentives would be adequate to convince Canada and the Province of British Columbia to move forward with "cooperative development" of the Columbia. The different perspec-

tives of Canadians and Americans regarding the negotiations that led
up to the Columbia River Treaty are reflected in two scholarly books
published shortly afterward by an American and a Canadian: John V.
Krutilla (1967); and Neil Alexander Swainson (1979).

In the midst of these negotiations, a new pro-development regime was
elected in British Columbia in 1952, catapulting to leadership W. A. C.
Bennett, who subsequently advocated grandiose hydroelectric development schemes for British Columbia's rivers. Realizing that Fraser River
development would be controversial, Bennett focused his attention on the
Peace River and Columbia River. Initially an advocate of private sector
development, Bennett turned to New Deal–style government-sponsored
megadam development after the privately owned BC Electric Company
balked at his plans for full utilization of British Columbia's rivers as a
hydroelectric cash register for provincial development. His influence
transformed international negotiations and increased the stakes for fishery interests. Both commercial and state fisheries interests in British Columbia had vigorously participated in decision-making regarding river
development for decades. Through their influence, political tradeoffs had
led to some rivers being developed and others protected. When Bennett
came to power there were as yet no dams on the main stem of the Fraser,
but five major proposals were pending in the 1950s. With Bennett's advocacy for full development, the fate of the Fraser was at stake. A "Fish
vs. Power" debate raged on both sides of the international border, with
fishery agencies, the canning industry, the International Pacific Salmon
Fishery Commission, and fishermen from both nations united in an effort
to keep dams off the Fraser. (The Pacific Salmon Treaty had allocated half
the Fraser's sockeye salmon run to salmon fishers from the United States,
so Americans were keenly interested in the fate of the Fraser, too.)

In contrast, hydropower interests in British Columbia were faction-
alized into public-versus-private power partisans, a circumstance that
prevented them from speaking with a united voice. The government of
British Columbia was spread thin boosting development simultaneously
all across the province, while the Canadian federal government was
mainly interested in the Columbia. As a consequence, fish advocates actually enjoyed some political leverage in this debate. As the historian
Matthew Evenden argues, Canadian authorities followed the path of
least resistance and concentrated their hydropower development efforts
on the Columbia and Peace Rivers. By 1958 a historic trade-off occurred,
and construction of dams on the Fraser River was off the table (Evenden

2004, 267–276). In the ensuing decades, the Columbia would be developed for hydropower and the Fraser reserved for salmon.

This tradeoff accelerated negotiations between the United States and Canada, leading to the signing of the Columbia River Treaty in 1961, which was subsequently revised and finally ratified in 1964. Among its agreements: Canada would build three large dams in the upper Columbia watershed and allow the United States to build a dam south of the border (Libby) that would back up water into Canada. The United States would pay Canada for one-half the downstream flood-control benefits plus one-half the downstream power-generation benefits that it derived from Canadian water storage. The deal-clincher was that the United States would pay Canada up front a lump sum for the first thirty years of hydropower benefits from Canadian storage of Columbia River water. Canada would then use that money to build the proposed dams (Krutilla 1967; Swainson 1979).

As a consequence, the Columbia River was fully developed for the most efficient hydropower system performance, including some uneconomic and highly destructive projects such as Libby and Dworshak dams (both United States "storage" projects), while the Fraser River remained dam-free. Virtually every remaining salmon run in the Columbia Basin is threatened or endangered, while the Fraser fishery remains healthy today—a symbol of the path not taken in the United States (Evenden 2004, 267–76; Lichatowich 1999, 170–201).

River development peaked in the United States in the 1960s and peaked in Canada in the 1980s (Harvey 2001). By the 1990s, officials on both sides of the border acknowledged that the great era of dam-building had ended. Electricity demand had outstripped hydropower supply starting in the 1960s. To compensate, the region's utilities were building more thermal power plants (coal, nuclear, natural gas) and jealously guarding their hydropower supplies. The politics of energy abundance in the Northwest had ended. Cities, energy-intensive industries, and irrigation farmers in competition over increasingly scarce water and expensive electrons aggressively advocated for their separate interests and sought to foist higher electricity rates on their competitors. Ignored in all this debate over electricity, fishery interests in the United States grew more desperate and vocal, but remained nonetheless disenfranchised from the hydropower system decision arena. Conflicts accordingly burgeoned in the 1970s, exacerbated by the passage of the Endangered Species Act in 1973 and a spate of significant legal victories for Northwest Indian tribes seeking to reassert their treaty-guaranteed fishing rights.

The U.S. Congress tried to address this conflict and empower all the affected interests to negotiate equitably with each other by enacting the 1980 Pacific Northwest Electric Power Planning and Conservation Act (Northwest Power Act). This odd package of compromises sought to guarantee cheap and abundant electricity in the region while simultaneously restoring endangered salmon runs. It required the Bonneville Power Administration as Northwest energy czar to equitably balance electricity needs with fish and wildlife conservation needs and to work with other federal, state, and tribal agencies to rebuild salmon runs on the Columbia River. Of course, such laudable goals were more easily mandated than implemented. Nevertheless, the Northwest Power Act represented the first significant step toward bringing greater equity and public participation into the decision arena and addressing the long-term legacy of disfranchisement and lopsided development in the Columbia Basin. So far, the Canadian path to resolving these conflicts seems to have produced more successful results (Blumm 1991; 1995; 1997).

In the current era, dam removal and river-restoration proposals fill newspapers and government reports the way dam-boosterism formerly did.[27] Recent studies of the Columbia River underscore the Northwest's ambivalence about its transformation. While there are still monographs with romantic titles like Paul Pitzer's *Grand Coulee: Harnessing a Dream* (1994), most of the literature is more critical of the Columbia River regime that has been built in the past century and its social and environmental costs. One of the more ambivalent studies is the historian Richard White's concise 1996 book, *The Organic Machine: The Remaking of the Columbia River*. Other writers have expressed a much more explicit lament. The *New York Times* reporter and award-winning author Blaine Harden titled his 1996 book *A River Lost: The Life and Death of the Columbia;* historian Keith Peterson's 1995 history of the lower Snake River is bluntly titled *River of Life, Channel of Death;* the salmon biologist Jim Lichatowich wrote provocatively in 2001 of *Salmon Without Rivers;* and historian Katrine Barber profiles the dispossession of Columbia River Indians in her 2005 book *Death of Celilo Falls*. Regret has replaced boosterism; contention has tarnished the cornucopian dreams of the river developers.

The apparent consensus for full development of the Columbia River, particularly in the United States but also in Canada after the 1950s, was built on a foundation of persistent inequities and the marginalization of dissenting voices, both of which returned to haunt river management regimes in the closing decades of the twentieth century. The people that the

Canadian authors J. W. Wilson and Tina Loo sympathetically referred to as being "in the way" (Wilson 1973; Loo 2004) have finally gained a voice, a place at the table, and a little power. In a limited, incremental fashion, a political transformation has occurred over the last generation. A series of statutes and court decisions in the United States have expanded public participation in decision-making, promoted values diversity, increased protections for the environment, and begun to redress historical injustices.[28] But the social inequities and environmental problems that accumulated for so many decades will not be redressed quickly or easily. Moreover, solutions to resource allocation problems in an age of scarcity will be much harder to craft and implement than solutions in an age of abundance.

Conclusion

The historical record reveals several interesting long-term cyclical trends in river development and policy in the Northwestern United States and Western Canada. One interesting long-term trend has been a semicyclical evolution from community-based decision-making responsive to broad social values for rivers in the early nineteenth century, toward market-based decisions by economic elites in the late nineteenth century, followed by state-based decision-making by political elites in the early- to mid-twentieth century, then back toward market-based decisions in the late twentieth century, with some movement most recently toward regional and community-based decision-making that is once again more responsive to broader social values and equity considerations. (On this recent trend toward broader values representation and participatory decision-making, see chapter 9.) Achieving sustainable and equitable river management seems more likely in the current regime than in previous periods since the rise of industrial capitalism.

There has been a cultural-ideological shift from a nineteenth-century view of rivers as integral, living, complex, and uncontrollable components of the human/natural landscape to a twentieth-century view of rivers as instruments for commerce and wealth, whose utility should be maximized through firm control (dams, dikes, and diversions), and finally—in recent decades—back to a more "ecological" view of rivers as only partially controllable living systems offering a variety of "ecosystem services" and sustaining diverse cultural values in addition to direct economic uses.

Several interesting trends in relationships have occurred over the past 150 years. Bilateral relations between the United States and Canada regarding water, hydropower, and salmon have not evolved in a cyclical fashion like the trends mentioned above, but rather have moved steadily from hostility, distrust, and isolation in the nineteenth century to tentative cooperation in the early twentieth century to increasing integration in the late twentieth century. Much of this trend toward cooperation and freer cross-border resource flows is attributable to the increasingly equitable nature of bilateral negotiations between the two nations. In contrast, intranational relations between rich and poor, between urban and rural, and between Native peoples and Euroamericans in the region have been more or less continually marked by distrust and conflict, attributable in large part to the very asymmetrical power relations and inequities between these groups.

Whether the environmental politics of a region are based on perceptions of abundance or of scarcity, resource allocations and water management regimes will succeed or fail based on how well they achieve social goals. As Helen Ingram has consistently argued, water is not simply an economic commodity: it is a public good. Its value goes far beyond market value. Because water is essential to both survival and social welfare, its protection and allocation are a matter of public interest and appropriately negotiated through the political process. Persistent inefficiencies and persistent inequalities usually return to trouble societies that permit them. While capitalist economics appropriately seeks efficiency as a goal in water use, democratic politics appropriately seeks fairness; and the two must be blended and balanced. In Ingram's words, "No lasting settlement of water allocations . . . is likely to be built on perpetuated inequity" (Ingram et al. 1986, 193). To achieve equity, she argues, requires sharing costs and benefits fairly, accommodating a plurality of values, establishing a widely inclusive political process, honoring social contracts and commitments, and protecting the needs of future generations.

This case study of rivers, hydropower, and salmon in the Northwest shows that river managers and developers violated each of these equity principles for much of the last 150 years, especially in the United States. Not surprisingly, those perpetuated inequities caused a great deal of social and political upheaval beginning in the 1960s and accelerating in the 1980s. Redressing historic injustices and restructuring water allocation processes to achieve greater participation and equity has contributed to the upheaval as beneficiaries of the old status quo resist surrendering their

historic advantages. But, as Ingram reminds us, those who have historically benefited from river development owe compensation to those who
have been historically marginalized or disenfranchised by that development (Ingram, et al. 1986). Achieving equity will not be easy or painless,
but it will, it is hoped, lead to a more socially acceptable and sustainable river-management regime. At the very least, we will have taken one
more step toward Wallace Stegner's hopeful dream of creating a society
to match our scenery.

Notes

1. Another leading historian, David Potter (1954), applied a similar thesis to an
analysis of American society as a whole in his influential book *People of Plenty:
Economic Abundance and the American Character.*

2. In addition to Mumme's chapter in this book, the history of water allocation conflicts and political negotiations over the Colorado River is concisely recounted by Norris Hundley Jr. (1986) in a chapter aptly titled, "The West Against
Itself."

3. My figures for the number of recognized US tribes and Canadian First Nations
in the Columbia and Fraser River Basins is drawn from maps at the following
two web sites: http://www.bced.gov.bc.ca/abed/map.htm (hosted by the government of British Columbia, Ministry of Education); and http://www.cr.nps.gov/
nagpra/DOCUMENTS/ClaimsMapIndex.htm (hosted by NAGPRA, National
Park Service, US Dept of Interior). I included several Puget Sound and Olympic Peninsula tribes in my count, despite the fact that they do not lie within the
boundaries of either river's watershed, because of their proximity to and historical use of Fraser River sockeye runs.

4. For a variety of perspectives and lots of data on the costs and results of salmon
recovery efforts, see United States Senate, *Bonneville Power Administration's annual fish and wildlife budget* (1998) and United States Congress, *Hydropower
river management and salmon recovery issues on the Columbia/Snake River system* (2001).

5. For a view of the complex social and political factors that go into US-Canada
transboundary relationships over salmon, see Sara Singleton (1998). For essays
on US-Canada relationships over water resources, see Robert Lecker (1991); and
International Transboundary Resources Center (1992).

6. The classic works on the relationship between Native peoples and Europeans
in the Northwest during the fur-trade era include Robin Fisher (1977); Sylvia Van
Kirk (1980); Thomas Vaughan (Vaughan and Holm 1982); and R. Cole Harris
(2002).

7. The phrase "legacy of conquest" was popularized by an influential history
of the American West by Patricia Nelson Limerick (1987); see especially Limer-

ick's 5, "The Meeting Ground of Past and Present," and 6, "The Persistence of Natives."

8. The Stevens Treaties are recounted in depth in a sympathetic biography of the governor by Kent D. Richards, *Isaac I. Stevens: Young man in a hurry.*

9. For a concise history of the past hundred years of laws and controversies over Northwest Indian fishing rights, see Charles F. Wilkinson (1992), chapter 5, "The River was crowded with salmon." For a more detailed account, see Michael Blumm (2002).

10. On the role of salmon in the subsistence economy of coastal and interior Northwest Indians, see Douglas Deur (1999) and Eugene S. Hunn (1999). See also, Joseph Taylor (1999), 1, and Jim Lichatowich (1999), 2. For a more thorough account of the traditional life-ways of Sahaptin-speaking peoples along the mid-Columbia River, see Eugene S. Hunn and James Selam (1990).

11. On the history of the Indian fishery on the Columbia River, see the citations in note 10 as well as Courtland L. Smith (1979), 2. On Celilo Falls specifically, see Katrine Barber (2005).

12. An interesting collection of case studies revealing the darker side of US Indian policy covering a wide range of topics in twentieth-century history is Fremont J. Lyden and Lyman Howard Legters (1992). For a survey of the ways in which Indian resources have been exploited largely for the benefit of non-Indians, see Donald Fixico (1998).

13. This provision is widely cited in histories of fish and wildlife law; e.g., Oregon Department of Fish and Wildlife (http://www.dfw.state.or.us/springfield/history_alt.html)

14. Ibid.

15. A fascinating social history of Pacific coast salmon canneries, which traces its northward migration, is Chris Friday (1994).

16. For the most detailed and sophisticated historical/legal study of this tragedy of the commons in Pacific fisheries, see Arthur McEvoy (1986). On the Northwest salmon fishery, the classic works on the topic are Anthony Netboy (1958; 1980). The best of the more recent studies of the salmon crisis include Evenden (2004), Lichatowich (1999), Joseph Taylor (1999), and Joseph Cone (1995).

17. For a quick survey and good references on economic activities that harmed salmon habitat in the nineteenth and twentieth centuries, see Goble and Hirt (1999, 236–38). For a more thorough historical analysis, see Lichatowich (1999), chapters 3 and 4.

18. An interesting cultural study of mining pollution and the conflicts between miners and the downstream farmers harmed by mining wastes that washed into the Coeur d'Alene River is found in Katherine Morrissey (1997).

19. On the history of water rights in the West and the doctrine of prior appropriation, see Donald J. Pisani (1996), chapters 1–3.

20. On the similarities and differences in US and Canadian western water law, see John McLaren (1992). On the similarities and differences between the US and

Canada in the development of their respective wests, see the chapter by Donald Worster (1998: 71–91).

21. For a fascinating analysis of the history and ecology of irrigated agriculture in the Snake River plain of southern Idaho, see Mark Fiege (1999). The marginalization of fish interests in the era of industrialization is recounted in a number of historical case studies. On the Pacific Northwest, see Jeff Crane (2004). On New England, see Theodore Steinberg (1991).

22. Much has been written about David Sohappy and his legal battle for Native fishing rights. Especially revealing is the award-winning documentary film titled "River People: Behind the Case of David Sohappy," Produced in 1990 by Michal Conford and Michele Zaccheo. See also the Columbia River Inter-Tribal Fish Commission's brief summary at http://www.critfc.org/text/usvor.html. A profile of Sohappy is also found in Bonnie Juettner (2005).

23. On the 1999 Pacific Salmon Treaty, see Daniel A. Waldeck et al. (1999); and Gordon R. Munro et al. (1998). On an earlier incarnation of the treaty, see M. P. Shepard and A. W. Argue (2005). US-Canada salmon treaty conventions have been revised regularly, with congressional hearings usually associated with them.

24. A great deal has been written about hatcheries and fish passage facilities at dams. For an illuminating historical analysis of the turn of the century effort to establish hatcheries, ensure dams on salmon streams had adequate fish ladders, and enforce other fishery laws, see Crane (2004), chapters on the Elwha River; Taylor (1999), 3, "Inventing a Panacea"; and especially Lichatowich (1999), chapters 5–6. For more recent studies of the fish passage problem, see the US government's "Annual fish passage report" for Columbia River Projects, Oregon and Washington, starting in 1968.

25. Statistics on the affects of dams on the Indian fishery are taken from the Center for Columbia River History Web site (Washington State University and Portland State University consortium), essay titled "Dams of the Columbia Basin and Their Effects on the Native Fishery," accessed at: http://www.ccrh.org/comm/river/dams5.htm.

26. "An assessment of Lower Snake River hydrosystem alternatives on survival and recovery of Snake River salmonids, appendix to the U.S. Army Corps of Engineers' Lower Snake River juvenile salmonid migration feasibility study," in Northwest Fisheries Science Center, available at http://purl.access.gpo.gov/GPO/LPS12311; also see United States Army Corps of Engineers (2002).

27. On dam removal as a recent political movement, see Elizabeth Grossman (2002). On dam removal news, see, for example, (Barnard 2006, C7; Durbin 2005, C1; *Columbian* 2004, C10).

28. A few of the policy innovations in the US that have dramatically transformed the sociopolitical environment of river management include: the National Wild and Scenic Rivers Act (1968), which provides a framework for designating river segments off-limits to development; the National Environmental Policy Act (1969), which requires environmental impact assessments and public par-

ticipation in federal agency decision-making; the Endangered Species Act (1973), which commits the federal government to recover species threatened with extinction; *United States v Washington* (the "Boldt Decision" of 1974), which assigned to Northwest Indian tribes half of the annual salmon harvest; and the Northwest Power Act (1980).

References

Barber, Katrine. 2005. *Death of Celilo Falls*. Seattle: University of Washington.

Barnard, Jeff (AP). 2006. "End in Sight for Dams that No Longer Make Sense." *The Columbian*. Vancouver, WA. January 21, p. C7.

Blumm, Michael C. 2002. *Sacrificing the Salmon: A Legal and Policy History of the Decline of Columbia Basin Salmon*. Den Bosch, The Netherlands: Book World Publications.

Blumm, Michael C. 1997. "Beyond the Parity Promise: Struggling to Save Columbia Basin Salmon in the Mid-1990s." *Environmental Law* 27, no. 1:21–126.

Blumm, Michael C. 1995. "Columbia Basin Salmon and the Courts: Reviving the Parity Promise." *Environmental Law* 25, no. 2:351–364.

Blumm, Michael C. 1991. "The Unraveling of the Parity Promise: Hydropower, Salmon, and Endangered Species in the Columbia Basin." *Environmental Law* 21, no. 3:657–744.

Blumm, Michael C. 1981. "Hydropower vs. Salmon: The Struggle of the Pacific Northwest's Anadromous Fish Resources for a Peaceful Coexistence with the Federal Columbia River Power System. *Environmental Law* 11 no. 43:211–261.

Bonneville Power Administration. "Multi-Purpose Dams of the Pacific Northwest." nd. Bonneville Power Administration, http://www.bpa.gov/Power/PL/Columbia/2-multi.htm (accessed 9/15/06).

Bonneville Power Administration. "Power Generation." nd. Bonneville Power Administration, http://www.bpa.gov/Power/PL/Columbia/2-gener.htm (accessed 9/15/06).

Brooks, Karl Boyd. 2006. *Public Power, Private Dams: The Hells Canyon High Dam Controversy*. Seattle: University of Washington Press.

Center for Columbia River History. Nd. "Dams of the Columbia Basin and Their Effects on the Native Fishery," Washington State University and Portland State University consortium. http://www.ccrh.org/comm/river/dams5.htm (accessed on 9/15/06).

Chacko, Chirakaikaran Joseph. 1968. *The International Joint Commission between the United States of America and the Dominion of Canada*. New York: AMS Press.

Columbia River Inter-Tribal Fish Commission. 2006. Columbia River Inter-Tribal Fish Commission: *U.S. v. Oregon*. http://www.critfc.org/text/usvor.html (accessed 9/15/06).

Columbian (Vancouver, WA). 2004. "Two Major Dams to be Removed on Olympic Peninsula." August 7, C10.

Cone, Joseph. 1995. *A Common Fate: Endangered Salmon and the People of the Pacific Northwest.* New York: H. Holt.

Conford, Michael and Michele Zaccheo, prods. 1990. "River People: Behind the Case of David Sohappy."

Crane, Jeff. 2004. *Finding the River: The Destruction and Restoration of the Kennebec and Elwha Rivers.* Pullman: Washington State University, Ph.D. dissertation.

Deur, Douglas. 1999. "Salmon, Sedentism, and Cultivation: Toward an Environmental Prehistory of the Northwest Coast," in Dale D. Goble and Paul W. Hirt, eds. *Northwest Lands, Northwest Peoples: Readings In Environmental History.* Seattle: University of Washington Press, pp. 129–155.

Durbin, Kathie. 2005. "Hemlock Dam to Be Razed to Aid Fish." *The Columbian* (Vancouver, WA), December 8, p. C1.

Evenden, Matthew D. 2004. *Fish Versus Power: An Environmental History of the Fraser River.* Cambridge, UK: Cambridge University Press.

Fiege, Mark. 1999. *Irrigated Eden: The Making of an Agricultural Landscape in the American West.* Seattle: University of Washington Press.

Fisher, Robin. 1977. *Contact and Conflict: Indian-European Relations in British Columbia, 1774–1890.* Vancouver: University of British Columbia Press.

Fixico, Donald. 1998. *The Invasion of Indian Country in the Twentieth Century: American Capitalism and Tribal Natural Resources.* Niwot: University Press of Colorado.

Friday, Chris. 1994. *Organizing Asian American Labor: The Pacific Coast Canned-salmon Industry, 1870–1942.* Philadelphia: Temple University Press.

Goble, Dale D., and Paul W. Hirt, eds. 1999. *Northwest Lands, Northwest Peoples: Readings in Environmental History.* Seattle: University of Washington Press.

Grossman, Elizabeth. 2002. *Watershed: The Undamming of America.* New York: Counterpoint.

Harris, R. Cole. 2002. *Making Native Space: Colonialism, Resistance, and Reserves in British Columbia.*

Harvey, Mark. 2001. "The Changing Fortunes of the Big Dam Era in the American West," in Char Miller, ed. *Fluid Arguments: Five Centuries of Western Water Conflict.* Tucson: University of Arizona Press.

Hundley, Jr., Norris C. 1986. "The West Against Itself: The Colorado River— An Institutional History," in Gary D. Weatherford and F. Lee Brown, eds. *New Courses for the Colorado River,* Albuquerque: New Mexico University Press, pp. 9–49.

Hunn, Eugene S. 1999. "Mobility as a Factor Limiting Resource Use on the Columbia Plateau" in Dale D. Goble and Paul W. Hirt, eds. *Northwest lands,*

Northwest Peoples: Readings in Environmental History. Seattle: University of Washington Press, pp. 156–172.

Hunn, Eugene S., and James Selam. 1990. *Nch'i-wána, "The Big River": Mid-Columbia Indians and Their Land.* Seattle: University of Washington Press.

Ingram, Helen M., Lawrence A. Scaff, and Leslie Silko. 1986. "Replacing Confusion with Equity: Alternatives for Water Policy in the Colorado River Basin." In Gary D. Weatherford and F. Lee Brown, eds., *New Courses for the Colorado River: Major Issues for the Next Century.* Albuquerque: University of New Mexico Press, pp. 177–199.

International Transboundary Resources Center. 1992. *Borders and Water: North American Water Issues: Papers from a Forum Held in Santa Fe, New Mexico,* June 26–27, 1992. Albuquerque, NM.

Isabella, Jude. 1999. "A Turbulent Industry: Fishing in British Columbia," reprinted from *Journal of the Maritime Museum of British Columbia* 45 (Spring), <http://www.goldseal.ca/wildsalmon/salmon_history.asp?article=3>.

Juettner, Bonnie. 2005. *100 Native Americans Who Changed History.* Milwaukee, WI: World Almanac Library.

Krutilla, John V. 1967. *The Columbia River Treaty: The Economics of an International River Basin Development.* Baltimore, MD: Published for Resources for the Future by Johns Hopkins Press.

Landeen, Dan and Allen Pinkham. 1999. *Salmon and His People: Fish and Fishing in Nez Perce Culture.* Lewiston, Idaho: Confluence Press.

Lecker, Robert. 1991. *Borderlands: Essays in Canadian-American Relations.* Markham, Ont: ECW Press.

Lichatowich, Jim. 1999. *Salmon Without Rivers: A History of the Pacific Salmon Crisis.* Washington, DC: Island Press.

Limerick, Patricia Nelson. 1987. *The Legacy of Conquest: The Unbroken Past of the American West.* New York: WW Norton.

Loo, Tina. 2004. "People in the Way: Modernity, Environment, and Society on the Arrow Lakes." *BC Studies,* no. 142–143:161–196.

Lyden, Fremont J., and Lyman Howard Legters. 1992. *Native Americans and Public Policy.* Pittsburgh, PA: University of Pittsburgh Press.

Martin, Irene. 1994. *Legacy and Testament: The Story of Columbia River Gillnetters.* Pullman: Washington State University Press.

McEvoy, Arthur. 1986. *The Fisherman's Problem: Ecology and Law in the California Fisheries.* Cambridge, UK: Cambridge University Press.

McLaren, John. 1992. "Water Law of the Canadian West: Influences from the Western States." In John McLaren, Hamar Foster, and Chet Orloff, eds. *Law for the Elephant, Law for the Beaver: Essays in the Legal History of the North American West.* Regina, Sask.: Canadian Plains Research Center.

Melosi, Martin. 1985. *Coping with Abundance: Energy and Environment in Industrial America.* Philadelphia: Temple University Press.

Morrissey, Katherine. 1997. *Mental Territories: Mapping the Inland Empire.* Ithaca, NY: Cornell University Press.

Munro, Gordon R., et al. 1998. *Transboundary Fishery Resources and the Canada-United States Pacific Salmon Treaty.* Orono: Canadian-American Center, University of Maine.

National Park Service, US Department of Interior. "Indian Land Areas Judicially Established 1978: Map Index." National Native American Graves Protection and Repatriation (NAGPRA), http://www.cr.nps.gov/nagpra/DOCUMENTS/Claims-MapIndex.htm (accessed 9/15/06).

Netboy, Anthony. 1980. *The Columbia River Salmon and Steelhead Trout: Their Fight for Survival.* Seattle: University of Washington Press.

Netboy, Anthony. 1958. *Salmon of the Pacific Northwest: Fish vs. Dams.* Portland, OR: Binfords and Mort.

Northwest Fisheries Science Center. 1999. "An Assessment of Lower Snake River Hydrosystem Alternatives on Survival and Recovery of Snake River Salmonids, Appendix to the U.S. Army Corps of Engineers' Lower Snake River Juvenile Salmonid Migration Feasibility Study," Seattle, WA: National Marine Fisheries Service, National Oceanic and Atmospheric Administration, http://purl.access.gpo.gov/GPO/LPS12311 (accessed 9/15/06).

Northwest Power Planning Council (US). 1986. *Staff Issue Paper: Hydropower Responsibility for Salmon and Steelhead Losses in the Columbia River Basin.* Portland, OR: Northwest Power Planning Council.

Norwood, Gus. 1981. *Columbia River Power for the People: A History of Policies of the Bonneville Power Administration, DOE-BP–7ed.* Portland, OR: Bonneville Power Administration.

Oregon Department of Fish and Wildlife. "Oregon's History of Fish and Wildlife Administration" Oregon Department of Fish and Wildlife, The South Willamette Watershed District Web Site, http://www.dfw.state.or.us/springfield/history_alt.html (accessed 9/15/06).

Pisani, Donald J. 1996. *Water, Land, and Law in the West: The Limits of Public Policy, 1850–1920.* Lawrence: University Press of Kansas.

Pitzer, Paul. 1994. *Grand Coulee: Harnessing a Dream.* Pullman: Washington State University Press, pp. 223–230.

Potter, David. 1954. *People of Plenty: Economic Abundance and the American Character.* Chicago: University of Chicago Press.

Province of British Columbia, Ministry of Education. 2001. "First Nations Peoples of British Columbia," <http://www.bced.gov.bc.ca/abed/map.htm>.

Richards, Kent D. 1979. *Isaac I. Stevens: Young Man in a Hurry.* Provo, Utah: Brigham Young University Press.

Robbins, William G. 1997. *Landscapes of Promise: The Oregon Story, 1800–1940.* Seattle: University of Washington Press.

Shepard, M. P., and A. W. Argue. 2005. *The 1985 Pacific Salmon Treaty: Sharing Conservation Burdens and Benefits.* Vancouver: UBC Press.

Singleton, Sara 1998. *Constructing Cooperation: The Evolution of Institutions of Comanagement.* Ann Arbor: University of Michigan Press.

Smith, Courtland L. 1979. *Salmon Fishers of the Columbia.* Corvallis: Oregon State University Press.

Stadfeld, Bruce Colin. 2004. *Electric Space Social and Natural Transformations in British Columbia's Hydroelectric Industry to World War II.* Ottawa: National Library of Canada.

Steinberg, Theodore. 1991. *Nature Incorporated: Industrialization and the Waters of New England.* Cambridge, UK: Cambridge University Press.

Swainson, Neil Alexander. 1979. *Conflict over the Columbia: The Canadian Background to an Historic Treaty.* Montreal: Institute of Public Administration, McGill-Queen's University Press.

Taylor, Joseph. 1999. *Making Salmon: An Environmental History of the Northwest Fisheries Crisis.* Seattle: University of Washington Press.

Tollefson, Gene. 1987. *BPA and the Struggle for Power at Cost.* Portland, OR: Bonneville Power Administration.

Ulrich, Roberta. 1999. *Empty Nets: Indians, Dams, and the Columbia River.* Corvallis: Oregon State University Press.

United States Army Corps of Engineers. 2002. *Improving Salmon Passage: Final: Lower Snake River Juvenile Salmon Migration Feasibility Report/Environmental Impact Statement: Summary,* Walla Walla, WA: US Army Corp of Engineers, Walla Walla District.

United States Congress, House Committee on Resources. 2001. Hydropower River Management and Salmon Recovery Issues on the Columbia/Snake River System: Field Hearing Before the Committee on Resources, House of Representatives, One Hundred Sixth Congress, second session, April 27, 2000, Pasco, Washington, Serial no. 106-94. Washington: U.S.G.P.O.

USGS (United States Geological Survey). "Description: Columbia River Basin, Washington," U.S. Department of the Interior, U.S. Geological Survey. http://vulcan.wr.usgs.gov/Volcanoes/Washington/ColumbiaRiver/description_columbia _river.html (accessed 9/15/06).

USGS (United States Geological Survey). 2002. USGS/Cascades Volcano Observatory, Vancouver, Washington; Description: Columbia River Basin, Washington.

United States Senate, Committee on Energy and Natural Resources, Subcommittee on Water and Power. 1998. Bonneville Power Administration's Annual Fish and Wildlife Budget: Hearing Before the Subcommittee on Water and Power of the Committee on Energy and Natural Resources, One Hundred Fifth Congress, second session, on the implementation of the 1996 Gorton Amendment to the Northwest Power Act, Vancouver, WA, February 17, 1998, Washington: U.S.G.P.O.

Van Kirk, Sylvia 1980. *"Many Tender Ties": Women in Fur-Trade Society in Western Canada, 1670–1870.* Winnipeg, MB: Watson and Dwyer.

Vaughan, Thomas, and Bill Holm. 1982. *Soft Gold: The Fur Trade and Cultural Exchange on the Northwest Coast of America*. Portland: Oregon Historical Society.

Waldeck, Daniel A., et al. 1999. *The Pacific Salmon Treaty: The 1999 Agreement in Historical Perspective*. Washington, DC: Congressional Research Service, Library of Congress.

Wallowa County (OR) 2002. *Board of Commissioners and Nez Perce Tribe, Wallowa County Nez Perce Tribe Salmon Habitat Recovery Plan with Multi-Species Habitat Strategy*. Enterprise, OR: Wallowa County Planning Commission.

White, Richard. 1991. *The Middle Ground: Indians, Empires, and Republics in the Great Lakes Region, 1650–1815*. Cambridge, UK: Cambridge University Press.

Wilkinson, Charles F. 1992. *Crossing the Next Meridian: Land, Water, and the Future of the West*. Washington, DC: Island Press.

Wilson, J.W. 1973. *People in the Way: The Human Aspects of the Columbia River Project*. Toronto: University of Toronto Press.

Witt, Sandra Johnson. nd. "The Lasting Legacy of the Deep River Finns." http://sydaby.eget.net/emig/deep_river.htm (accessed 9/>15/06).

Worster, Donald. 1998. "Two Faces West." In Paul W. Hirt, ed., *Terra Pacifica: People and Place in the Northwest States and Western Canada*. Pullman: Washington State University Press, pp. 71–91.

II

Civic Engagement and Governance

One of the central themes of this book is that communities develop their governance capacity through collective action on water problems. The chapters in this section consider how water governance fares and has fared in terms of equity. According to Elwood Mead, an early observer of water development in the American Southwest, "There is either murder or suicide in the hearts of every member" of irrigation communities until equitable institutions for allocating water are developed (Maass and Anderson 1978, p. 2.). While water is a life-and-death matter in many contexts, it is also too important to fight over physically. Aaron Wolfe, who has made a thorough study of international water agreements and compacts, finds a multitude of negotiated settlements and institutional structures for governance of shared resources, but few wars have been fought over water (Wolf 2002). Communities since the dawn of civilization have managed to pull themselves together in order to deliver water. Roman authorities considered that the presence of an aqueduct and public fountain were signals of the arrival of culture (Ingram, Scaff, and Silko 1986).

Procedural equity requires open, transparent, accessible, and accountable decision making. The chapters in this section explore the current wave of water reform across different countries and find a renewed dedication toward making such processes a reality. They also document frequent failures. Contemporary water reformers place great faith in the capacity of water markets and decentralized watershed institutions to govern closer to the problem and to utilize more contextual understanding about, and more informed interest in, water choices. One assumption is that if water is priced at its true cost, the "stuff that comes out of the tap" will get people's attention more fully and receive more careful treatment. The world described by the chapters in this section reveal matters to be a great deal more complex.

Madeline Baer in chapter 7 examines the unraveling of a privatizing and rationalizing scheme to improve water management in a city in Bolivia. The Bolivian government and the international corporation that received the privatization business opportunity failed to inform and involve citizens in the process, and that mistake spelled disaster for their plans and for the people of the city of Cochabamba. Partisan and economic class differences within the community were shoved aside as citizens mobilized to reclaim public control over their water supply. It remains to be seen whether this capacity of citizens to demonstrate against privatization can be translated into equitable and effective urban water management. In this case, the citizens were left with a circumstance even worse than before the failed privatization.

Chapter 8 reminds the reader that water must be considered in the context of the postmodern contemporary concern with issues of lifestyle and choice. Ismael Vaccaro observes the transition from tradition to modernity in the central Catalan Pyrenees, and concludes that local residents have never been in charge of their own destiny. Transformations have followed the dictates of national and international forces beyond local control.

Chapter 9 considers technoscientific information about climate change and the availability of water and how that information might be used to better inform democratic decision-making processes about water problems. Maria Carmen Lemos looks at what happens when watershed councils in Brazil are presented with data on the probabilities of future drought. The findings are disappointing but not surprising. The decentralization of water resources decision-making power to the watershed level has not erased the power differences between ordinary farmers and technical elites. Equity issues need to be conducted in value terms. The tendency on the part of technical experts to objectify and rationalize water decisions marginalizes equity discussions and the nonelites that would be served by greater focus on equity.

The concluding chapter in the volume considers the equity implications of the most recent and most aggravating challenges to water resources. The competition among the values associated with water supply and management will increase in the year ahead requiring the redesign of institutions and policies so that equity issues are specifically addressed.

The changing climate guarantees increased competition among the values associated with water. Water conflict is inevitable. It cannot be avoided through any application of universal principles. Attention needs to shift to process, tools, venues, and institutions through which conflict can be dealt with while attending to equity.

References

Ingram, Helen, Lawrence Scaff, and Leslie Silko. 1986. "Replacing Confusion with Equity: Alternatives for Water Policy in the Colorado River Basin.," In G. D. Weatherford and F. L. Brown. *New Courses for the Colorado River: Major Issues of the Next Century.* Albuquerque: The University of New Mexico Press, pp. 177–199. [Reprinted in J. Finkhouse and M. Crawford, eds. 1991. *A River*

Too Far: The Past and Future of the Arid West. Nevada Humanities Committee. Reno: University of Nevada Press, pp. 83–102]

Maass, Arthur, and Raymond L Anderson. 1978. *And the Deserts Shall Rejoice: Conflict, Growth, and Justice in Arid Environments.* Cambridge, MA: MIT Press.

Wolf, Aaron T., ed. 2002. *Conflict Prevention and Resolution in Water Systems* Cheltenham: Edward Elgar. http://www.transboundarywaters.orst.edu/data/

7

The Global Water Crisis, Privatization, and the Bolivian Water War

Madeline Baer

The struggle for access to potable water is at the nexus of the larger battle between states, multinational corporations, international financial institutions, and organized groups of citizens in Latin America. The common terrain for these actors is the emerging water crisis, a result of the rapidly dwindling supply of available clean water for human consumption. Less than 3 percent of the world's water supply is fresh water, and less than 1 percent is accessible for human consumption. More than 1 billion people lack safe drinking water, and 2.6 billion people lack basic sanitation (World Health Organization 2004). Transnational water companies see a financial opportunity in this crisis, as limited supply and growing demand motivate the scramble to control the world's water systems. Environmentalists see an ecological crisis of grave proportions, which calls for an immediate transformation of the way water is extracted and used. On the community and individual level, many Latin American citizens have experienced limited access to clean water under both state-owned and privately owned water systems.

This chapter examines water privatization policy in Cochabamba, Bolivia, the site of the first large-scale rejection of water privatization in Latin America. The privatization of water emerged as a controversial policy area in Bolivia after massive protests forced the retreat of a multinational water conglomerate from the city of Cochabamba. The citizens of this Bolivian city mobilized to express the viewpoint that water is a human right and a natural resource that is too vital for the preservation of life to be treated solely as an economic good. In terms of the broader conceptualization of this book, the protests in Cochabamba are an example of how inequitable water management policy can cause social unrest and political instability.

The United Nations Commission on Sustainable Development describes water as "irreplaceable for the purposes of drinking, hygiene, food production, fisheries, industry, hydropower generation, navigation, recreation, and many other activities. Water is equally critical for the healthy functioning of nature, upon which human society is built" (UNCSD 1997, cited in Finger and Allouche 2002, 32). Safe drinking water and sanitation are directly linked to human health and environmental health; 80 percent of diseases in poor countries are spread through unsafe water, and an estimated 90 percent of Third World wastewater flows untreated into local rivers and streams (Barlow and Clarke 2002, 52).

Global water shortages are emerging as a result of overpopulation, overuse, pipe leakage, pollution, and waste. The poor suffer disproportionately from water shortages and lack of reliable access to safe water and sanitation services. Women bear the brunt of water scarcity and lack of water rights, as women play a central role in the provision and management of water in poor countries (Bennett, Davila-Poblete, and Rico 2005). As the World Bank notes, addressing the global water crisis is essential for reducing poverty in the developing world (World Bank 2004a).

In this chapter, I analyze water privatization policy in Bolivia using Schneider and Ingram's (1997) policy design for democracy approach, which asserts that policy designs have a direct impact on opportunities for democracy and citizenship. Through analysis of the policy design, water privatization policy in Bolivia is found to have had negative consequences for democracy and active citizen participation. Despite this negative impact on democratic citizen participation, a social movement emerged in Cochabamba that confronted and overturned the unpopular policy. I begin with a brief overview of World Bank water policy, privatization, and the circumstances leading to the reversal of the water contract in Cochabamba. Next, I analyze the content and design of water privatization policy in Cochabamba, focusing on the policy goals, policy implementation, and the social constructions of knowledge that were embedded in the policy. This analysis also explores the World Bank's role in pushing privatization in Bolivia, and raises concerns about accountability and the democratic deficit within the World Bank.

Decisions made at the supranational level by international financial institutions have a tremendous impact on local policies in developing countries. Economic reforms pushed by international financial institutions like the World Bank and the International Monetary Fund have met with varying degrees of resistance, particularly in Latin America, where many social movement groups and nongovernmental organizations protest the social and economic consequences of these reforms. Globalization processes are changing the role of the state in the everyday lives of citizens, and transnational corporations are playing an increasing role in determining "who-gets-what in the world system" (Strange 1996, 54). While international financial institutions and multinational corporations are pushing the state out of the economy and lessening its role, many Latin American social movements are simultaneously pulling the state back into local communities and demanding a strong government role in the provision of services. These social movements often resist public policies that lessen the role of the state in the daily lives of citizens.

Public policies have far-reaching and long-lasting impact beyond the specific outcomes that arise from their implementation. In addition to allocating resources, policies can reproduce and reify existing social constructions of groups, institutions, and knowledge (Schneider and Ingram 1997). In this way, public policies have the potential to strengthen democracy and galvanize active citizenship, or they can have a negative impact on opportunities for democratic participation. As a response to unpopular policies, social movements can emerge to protest and effect change in public policy.

When citizens challenge policies, they often focus on the negative social and economic outcomes of policies. Social movements also frequently challenge the democratic deficits that exist within the institutions involved in making and implementing unpopular policies. Water privatization is a policy area where claims have surfaced regarding negative social and economic effects as well as undemocratic decision-making processes by institutions like the World Bank.

World Bank Water Policy

The World Bank is the most influential actor in the international water sector, both in terms of policymaking and financial aid for water-related projects (Bauer 2004; Finger and Allouche 2002; Moore and Sklar 1998). The water sector represents 15 percent of the Bank's cumulative lending (World Bank 1993). Although Bank-financed water projects have experienced successes over the years, many water supply and sewerage projects financed by the World Bank have a poor track record in terms of efficiency, improved service, and improved access for the poor (Moore and Sklar 1998).[1]

The World Bank's approach to water policy drastically shifted in the early 1990s from a model that promoted state involvement in infrastructure development, to one that portrays the state as an impediment to progress, development, and the public interest (Finger and Allouche 2002). While the Bank had previously viewed infrastructure services as public goods warranting state provision, it began to advocate treating these services as private goods. This shift coincided with the rise of the "Washington Consensus" as the dominant paradigm for economic development. The Washington Consensus called for the privatization of public utilities such as electricity and water.

The World Bank's 1993 Water Resources Management Document elaborated this shift in policy goals to reflect the newly accepted free-market approach. It also incorporated concerns from environmental groups and various NGOs that were pushing the Bank to adopt more environmentally sustainable policies and to be more responsive to local community input. The new approach to managing water resources included treating water as an economic good, decentralizing management and delivery structures, placing a greater reliance on pricing, fuller participation by stakeholders, and a more comprehensive policy framework that considers social, economic, and environmental aspects of water management (World Bank 1993).

Proponents of water privatization argue that treating water like any other commodity is the only way to improve efficiency and address community water needs. Privatization supporters cite studies that show how privatization has benefited the poor and improved service in some developing countries (Chong and Lopez-de-Salinas 2004; Galiani et al. 2003). Those who oppose the commodification of water challenge the notion that transnational corporations have the right to own and profit from entire water systems, which include the delivery mechanisms for water, the underground aquifers, and water resources that are naturally replenished through the hydrologic cycle. Proponents of equitable water management often challenge the notion of efficiency as the appropriate criterion for water policymaking. As the debate over water privatization grows, two competing visions of water emerge: one that favors defining water as an economic good that should be commodified and traded on the open market; and the other that defines water as a public trust, a part of the global commons, and a basic human right that must not be privatized (Barlow and Clarke 2002).

Privatization in Latin America

Beginning in the early 1990s, privatization of water was viewed as the so-lution to water crises throughout Latin America. Public sector failure and the need for substantial investment coincided with a growing number of transnational corporations that were seeking to enter water markets (Finger and Allouche 2002). Ten corporations or consortiums currently dominate the global water industry, the two largest of which are Vivendi Universal and Suez (Barlow and Clarke 2002, 106). Smaller consortiums such as RWE-Thames and Bechtel-United Utilities also provide water to millions of people worldwide and are increasingly challenging Vivendi and Suez's dominance in the water sector (Barlow and Clarke 2002). As many governments in the developing world abandon their traditional roles in the provision of services like water and sanitation through the privatization of public utilities, these corporations have come to own and operate many Third World water systems.

The implementation of privatization of services in Latin America is at-tributed to a mixture of state inefficiency, changing norms about the role of the state in the economy, the lack of resources in some countries, and the acceptance of the market as an appropriate venue for providing ser-vices to citizens. Advocates for privatization of public services argue that by selling state-owned enterprises to the private sector, developing coun-tries eliminate problems of corruption and inefficiency that bog down state-run enterprises. Developing countries are increasingly under pres-sure to privatize public utilities such as telecommunications, electricity, and water in order to receive loans from international financial institu-tions such as the World Bank and the International Monetary Fund. This policy is challenged throughout the hemisphere by many labor unions, NGOs, and local community organizations with the support of trans-national advocacy groups. These sectors claim that privatization has a negative impact on access, quality, and accountability of social services.

Privatization in Bolivia

Bolivia began implementing neoliberalism in 1985 with a set of reforms called the "New Economic Policy." The New Economic Policy was a group of structural adjustment reforms that completely restructured the economic and social fabric of the country. It included reduction of the public sector, an end to state subsidies, wage freezes, and a general retrenchment of state agencies responsible for social welfare issues such as health and education (Gill 2000). Although the World Bank and the International Monetary Fund have consistently pressured Bolivia to adopt neoliberal reforms, initial acceptance of neoliberalism in Bolivia came from political leaders within the country. The central idea in the neoliberal approach is that free markets increase efficiency through pricing, while state regulations and subsidies distort this process and therefore reduce efficiency.

By 1985 Bolivia faced a growing national economic crisis. This crisis was the culmination of years of fiscal mismanagement by military regimes, which resulted in massive foreign debt, inflation, and a stagnated tin industry (formerly Bolivia's primary source of foreign exchange). There was a need for massive economic reform in Bolivia at this time. Members of the government who chose to pursue the neoliberal route found strong allies in the International Monetary Fund and the World Bank (Gill 2000).

Under pressure from the World Bank and the International Monetary Fund, the Bolivian government began widespread privatization of state-owned industries including the national airline, railroads, tin mines, the telecommunications industry, and the national electric company (Shultz 2003b; Olivera 2004). The World Bank began aggressively pushing for privatization of Cochabamba's public water system in 1996. The Bolivian government finally negotiated a contract for the sale of the system in 1999. The contract for Cochabamba's water system was the first water privatization in Latin America to be subsequently reversed in the face of massive social protest. Although the World Bank touted the policy as the solution to the city's water problems, the privatization of water delivery services in Cochabamba quickly became a political quagmire for the Bolivian government and a symbol of resistance to neoliberal reforms in Latin America.

Cochabamba's Water War

The Bolivian "water war" of April 2000 was the culmination of years of domestic unrest over water issues in Cochabamba (Assies 2003). Prior to privatization, Cochabamba suffered chronic water shortages and many of the poorest neighborhoods were not connected to the municipal water network. Under this arrangement, state subsidies to the water utility mainly supported middle-class neighborhoods and industries located in the urban center (Finnegan 2002). Communities that were not connected to the urban network had to seek alternative sources of water. Many of these marginalized communities pooled their resources with international aid to create independent freshwater well cooperatives which provided clean and inexpensive water to members of the cooperative. These cooperatives served a relatively small number of people, while the majority of Cochabambans endured inadequate water quality and access.

Under pressure from the World Bank to accept the conditions of a debt-relief package (Shultz 2005a), the Bolivian government sold the Cochabamba water system in 1999 to the only bidder, Aguas del Tunari, a subsidiary of the United States corporation Bechtel.[2] The forty-year contract gave Aguas del Tunari exclusive control over the municipal system, including all industrial, agricultural, and residential systems, as well as the water in the natural aquifer. The deal also allowed the company to install meters on all water sources, including cooperative wells (Finnegan 2002).

These acts were legal under Law 2029 (the Drinking Water and Sanitation Law), which was approved by the Bolivian government in October 1999, in preparation for the privatization. Law 2029 was passed quickly with little public knowledge, deliberation, or public support (Olivera 2004; Slattery 2003). The new law created the legal framework for private ownership of water sources. It authorized private responsibility for delivery of water to citizens (Slattery 2003). The law outlawed traditional water practices, such as cooperative water systems and individual homeowners' wells, and banned collection tanks used by many peasants to collect rainwater. It ended government subsidies for water consumers and authorized price increases that would allow the price of water to reflect the "true economic cost of the service" as determined by the contracted owner (Slattery 2003).

The impact of the privatization was immediate and widely felt. The new water company installed meters on cooperative wells that were built with donated funds and the personal funds of the community. The communities were charged for water drawn from the wells and for the installation of the meters (Finnegan 2002). Some urban families experienced a 200–300 percent increase in water rates (Schultz 2003a). Lower-income workers faced water bills that exceeded a quarter of their monthly income (Finnegan 2002).[3] Farmers, engineers and environmentalists were outraged over the lack of consultation by the government with local experts in negotiating the contract (Finnegan 2002). Middle-class families and business owners objected to the rate hikes and to the lack of improvement in water quality under the new water company. In response, a coalition of neighborhood associations, labor unions, workers, farmers, and other sectors of society formed the Coalition for the Defense of Water and Life, or "La Coordinadora."[4]

Although La Coordinadora was officially formed in December 1999, some members of La Coordinadora were working on water management issues in Cochabamba for years prior to the privatization of the municipal system. The Cochabamba Department Federation of Irrigators' Organizations (FEDECOR) had been organizing and protesting around issues of resource ownership in the rural areas surrounding Cochabamba since the mid-1990s, focusing mainly on providing alternative proposals for water management to the government's proposed general water law (Assies 2003). The Society of Bolivian Engineers (SIB) had also challenged government proposals for water management in the area since the late 1990s. These groups protested the exclusionary and nontransparent negotiation of the contract between the government and Aguas del Tunari. The government rejected their requests for inclusion in the negotiation of the contract and design of the policy despite their interest and expertise in water resource management in Cochabamba (Assies 2003).

When the contract with Aguas del Tunari was signed in late 1999, the Cochabamba Department Federation of Irrigators' Organizations and the Cochabamba College of Engineers predicted that the new contract would mean a drastically reduced quantity of potable water for consumers and agriculturalists, as well as massive rate hikes that would disproportionately hurt the poor (Assies 2003). When the policy was put into effect in January 2000, the predictions were realized; Aguas del Tunari instituted massive rate hikes and carried out seizures of private and communitarian water systems. As the implementation of the policy progressed, trade unionists, community organizations, and consumers joined in opposition to the policy.

Although the protests that took place in the early months of 2000 were mainly over rate hikes and the seizure of private water sources, the leadership and organized roots of La Coordinadora consistently exposed the policy negotiation process between the government and Aguas del Tunari as undemocratic and nontransparent. Thus, the protests in Cochabamba reflected a mixture of normative complaints about the negotiation process made by organized communities and organizations such as

FEDECOR and SIB, as well as utilitarian concerns about cost and owner-ship made by consumers.

In the early months of 2000, La Coordinadora organized demonstra-tions and highway blockades that paralyzed most of the country. La Coordinadora demanded that the government review the rate increases enacted by Aguas del Tunari and renegotiate the contract with the com-pany. Dissatisfied with the government's response to the rate hikes, they eventually called for the government to modify Law 2029 and to cancel the contract with Aguas del Tunari. The protesters framed the issue as a fight between poor people and a greedy multinational corporation over the basic human right to water (Finnegan 2002).

In April 2000, more than one hundred thousand citizens from Cocha-bamba and the surrounding areas participated in a general strike and multiple highway blockades across the nation, which led the government to declare martial law. Conflicts between protesters and police left dozens wounded and one person dead. When it became clear that the protest-ers were too numerous and too angry to back down, Aguas del Tunari executives fled Cochabamba. Shortly after, the government revoked the contract with Aguas del Tunari. In a victory for La Coordinadora, Co-chabamba's water system was returned to the public utility with a new board of directors that included Coordinadora representatives.

A new national water law was written giving legal recognition to *usos y costumbres* (traditional communal practices), including protection of independent water systems, public consultations on rates, and a com-mitment to prioritizing social needs (Finnegan 2002). But with a sinking aquifer, crumbling infrastructure, and little chance of new international investment, the public system is desperately in need of capital. Bechtel at-tempted to sue the Bolivian government for over $25 million in lost prof-its in the World Bank's International Center for Settlement of Investment Disputes, but was forced to drop its suit due to international pressure.

The Cochabamba water war became a symbol of resistance to water privatization and of rejection of neoliberal policies that reduce the role of the state in domestic economies. Opponents of water privatization objected not only to the rate hikes, but also to the lack of democratic decision-making processes and consultation associated with the privatiza-tion (Olivera 2004; Assies 2003). Public statements by La Coordinadora decried the "simulation of democracy which only renders us obedient and impotent" (La Coordinadora 1999). The conflict over water privati-zation in Cochabamba was the first of a number of consumer rebellions in Bolivia against the sale of natural resources to private corporations (Shultz 2005b). Organized consumers reversed the privatization of water in the Bolivian city of El Alto in 2005. Ongoing calls for renationalization of the country's natural gas and oil reserves have created political instability since 2003 (Painter 2005), leading to the nationalization of the nation's oil and gas reserves in May 2006 (McDonnell 2006). The failure of water privatization in Cochabamba is an example of how public policy is more than the sum of its economic outcomes; it also has wider implications for democracy, citizenship, and for the prospects of equitable water management.

The Policy Design of Water Privatization in Bolivia

Schneider and Ingram (1997) provide a framework for analyzing policy designs and the dynamic relationship between policies and their contexts. "Policy design" refers to the instrumental and symbolic content of the public policy, such as the architecture, discourses, and practices that comprise the policy. Policy designs are created through a dynamic process that involves the social construction of knowledge and of target population identities. Policy designs have consequences for democracy and active citizenship; policy designs can enable and support active citizen participation and democracy. They can also contribute to the reproduction of degenerative policies that undermine and stifle democratic participation. Degenerative policies discourage active citizenship by excluding citizen participation, devaluing local knowledge and input, and are often handed down from above with little transparency and accountability. Conversely, policies can enable citizen participation by strengthening communicative processes through fostering communication among different levels of government and public and private actors (Schneider and Ingram 1997).

Policy designs emerge from specific issue contexts which are the socially constructed understandings that emerge from the broader societal context. These social constructions are then embedded in the design itself. By socially constructing an issue, a set of beliefs is accepted as "true" or "real." The policy reflects and reproduces this particular construction of reality. Unpacking a policy's design involves looking at this issue context as well as the core empirical elements of the policy. These core policy elements include observable elements, such as the goals or problems to be solved, agents, target populations, rules, tools, rationales and assumptions that make up the policy (Schneider and Ingram 1997).

In the case of water privatization in Bolivia, an examination of the policy goals, policy implementation, and social constructions embedded in the policy reveals a discrepancy between stated and hidden goals, an implementation structure that limited opportunities for democratic participation, and a policy rationale that reproduced socially constructed knowledge about water and about the appropriate method of delivering water to citizens.

World Bank Water Policy Goals

The goals of public policies correspond to the problems to be solved by the policy, or what is to be achieved or changed through the policy (Schneider and Ingram 1997, 82). Policy goals reflect conscious choices that are made about what the policy is intended to achieve, and they result in "benefits to some and burdens to others" (Schneider and Ingram 1997, 83). This allocation of benefits is linked to concerns about equity, justice, and democracy. Goals can be normative, political, and technical in nature. They reflect the societal context in which they are formed. Goals can also reflect a politically motivated rationale; these goals are often hidden, and can be revealed in examination of the implementation and outcomes of the policy.

The overall goal of the World Bank's water policies is to reduce global poverty through sustainable development. The World Bank's public goals regarding water management are linked to the stark reality of massively inadequate access to clean water in the developing world. Public World Bank documents refer to the millions of people, the majority of whom are extremely poor, who lack access to safe water and decent sanitation. The World Bank's goal is to facilitate the provision of effective, fair delivery of water and sanitation services in developing countries. The World Bank is committed to helping achieve the Millennium Development Goal of halving by 2015 the proportion of people without sustainable access to safe drinking water and basic sanitation (World Bank 2004b).

The World Bank asserts that "improving service efficiency is . . . an overarching goal for the reform of water and sanitation services" (World Bank 2005). Improving efficiency requires "investment coupled with reform" (World Bank 2002b). Proposed reforms involve "creating the conditions for more accountable, efficient, equitable and effective service providers" (World Bank 2002b). Achieving efficiency in the water sector requires reforms that allow private sector control of services under contract with the public sector. This type of reform calls for a transparent financial planning process to identify "the right sources of finance" to support the long-term goals (World Bank 2005). The World Bank had specific goals for water privatization in the three Bolivian cities of Cochabamba, La Paz-El Alto, and Santa Cruz de la Sierra. These goals were to "provide service for all in the most efficient manner" and to "lay the foundation for sustainability" (World Bank 2002a).

The World Bank favors water policies that "put people at the center of decision-making" and "at the center of sustainable water supply and sanitation" (World Bank 2002b). These policy goals emphasize that communities ought to be in control of their own water systems and they must be empowered to make decisions regarding water services. The World Bank encourages borrowing countries to implement reforms to allow this kind of community empowerment and control. The Bank is committed to facilitating participation of affected people and NGOs in water resource management, with an emphasis on the poor and indigenous peoples of borrowing countries (World Bank 1993). The rhetoric the World Bank uses to describe its policy goals is imbued with respect for local communities and local knowledge, support for empowerment of citizens, and an overarching humanitarian mission to improve the lives of the poor.

The World Bank has a declarative public commitment to serving the poor and involving local communities in decision making regarding water policy. There is evidence that the World Bank acted contrary to these goals in its involvement with Bolivia. The World Bank emphasizes that water policy must prioritize the needs of the poor. Yet prior to the negotiations for the privatization contract in Cochabamba, World Bank officials specifically informed the Bolivian government that "no public subsidies should be given to ameliorate the increase in water tariffs in Cochabamba" (World Bank 1999). Bolivian groups, such as the Cochabamba College of Engineers and FEDECOR, had predicted that post-privatization rate increases would leave many residents unable to afford the new price for water (Assies 2003). This was a reasonable prediction given that over 70 percent of the Bolivian population lives below the poverty line.

This rejection of subsidies contradicts many internal World Bank analyses regarding privatization. Some analysts within the Bank point to the necessity of these safeguards during the initial stages of privatization to protect the poor (Klein 2003; World Bank 1993: and see Slattery 2003 for a non–World Bank, pro-privatization analysis of the importance of subsidies). Certain services require a minimum level of consumption for survival, such as water and heating, and thus may require lifeline exemptions for some users who are too poor to afford market prices (World Bank 1993; Finger and Allouche 2002). The World Bank specifically advised the Bolivian government against this method of protecting the poor during the time when the Bank was assisting Bolivia with the design of the privatization policy (World Bank 2002a, 2).

The World Bank's goal of ensuring local control over water management was also contradicted by the World Bank's policy recommendations and demands on the Bolivian government. The concession to Aguas del Tunari essentially outlawed the local small-scale water systems, such as community wells and cooperative water systems that had been built with international aid and community funds. This concession contract, prepared "with the Bank's assistance" (World Bank 2002a, 2), gave sole control of all water delivery and sanitation services to the private contractor. This included the water in the underground aquifer, which residents drew from in their private wells. It also effectively privatized rainfall, as residents were no longer permitted to capture rain for their own consumption without paying the new water company. Aguas del Tunari now owned these methods of water supply that were primarily used by the poor. In practice, the policy did not meet the World Bank's goals of maintaining local control over water management and encouraging active citizenship.

Contrary to its stated goals of engaging in open and participatory processes with local communities to decide on appropriate reforms, the World Bank pressured the Bolivian government to adopt the policy by making privatization a condition for loans and debt relief. In February 1996, World Bank officials informed the mayor of Cochabamba that a pending $14 million loan to improve water service was conditioned on privatization of the city's system ("Banco Mundial Es Claro . . ." 1996, 10). Bank officials also pressured Bolivia's president to privatize the system in June 1997 by making $600 million in badly needed international debt relief contingent on the privatization of Cochabamba's water system ("Organismos multilaterales presionan al Gobierno . . ." 1997, 5).

Community representatives were not allowed to participate in the decision-making process over the privatization of water in Cochabamba.

There was no open and participatory process where local residents were invited to deliberate and decide on appropriate reforms to the water sector (Shultz 2005b; Olivera 2004). This dismissal of local input counters the World Bank's stated goals of community involvement and respect for local knowledge in decision making about water policy. The Bolivian government also bears responsibility for the lack of consultation with local groups. Contrary to claims by members of La Coordinadora, one Bolivian analyst asserts that consultation did take place, in the sense that local elected officials and representatives from the municipal water system (SEMAPA) were involved in the process of negotiation of the contract (Laserna 2000). However, most accounts of the process cite a total dismissal of local input from those affected by the policy (See Olivera 2004; Shultz 2003b and 2005b; Public Citizen 2003). Many argue that those who were consulted did not represent the needs of the public (Assies 2003).

Critics of World Bank water policies claim that its unwavering support for privatization of water in the developing world is rooted in ideological support for the neoliberal model of development. The World Bank's shift to water privatization policies in the 1990s mirrored the global trend toward rolling back the government's role in the economies of developing countries. Although the state had previously played a prominent role in World Bank–funded infrastructure projects, the World Bank began portraying the state as an impediment to progress. By reframing the government as inefficient in providing these services, the goals of privatization policy became linked with delegitimizing public control of utilities and loosening national control over domestic economies. This perspective describes the World Bank's goals as a reflection of the dominant ideological framework of the time; a framework that is pushed by western developed nations, such as the United States, that maintain significant influence within the World Bank.

Although the World Bank's public goals regarding water management call for policies that support democratic participation by local communities and active citizen roles in decision making and resource management, there is a disconnect between these goals and the resulting policy "on the ground" in Bolivia since 1996. Charges of hidden goals that support the financial interests of private multinational corporations and the ideological policy perspectives of western developed countries call into question the World Bank's role as an international organization primarily concerned with poverty alleviation and responsible resource management.

Imposing policies without consultation, removing local control over natural resources, and discouraging financial support for the poor in adjusting to privatization effectively removes local people from decision making regarding the policy, and prevents them from participating in the implementation of the policy.

Implementation of World Bank Water Policy in Bolivia

In Schneider and Ingram's policy design framework, implementation is defined as the "value added" to design by the agents who carry out the policy (Schneider and Ingram 1997, 89). Although the World Bank prescribed the policy, the Bolivian government and Aguas del Tunari were the main agents responsible for implementing the policy. Implementation of the policy began with the negotiation process to decide on the nature of the privatization agreement, and included the actual implementation of the contract in Cochabamba. The implementation of the policy and the corresponding messages sent to the public showed a disregard for democracy and citizen participation, as local input and needs were systematically ignored.

By most accounts, the negotiation process regarding the privatization of Cochabamba's water system was closed and undemocratic, with no citizen role in negotiation or bargaining. During the government negotiations with Aguas del Tunari, citizens attempted to express their concerns over the negotiation process. Local engineers and environmentalists wanted to participate, as did local farmers who feared their irrigation systems would be threatened. The president of FEDECOR, Omar Fernandez, expressed concerns that the contract with Aguas del Tunari would result in a rise in the price of irrigation water that would drive between 15,000 and 20,000 farmers into bankruptcy. The Ministry of Foreign Trade and Investment and the World Bank later set the prices at the level predicted by Fernandez (Assies 2003). Despite the expertise of groups like FEDECOR and the wealth of local knowledge about the needs of local farmers and consumers in Cochabamba, these sectors of the community were not incorporated into the negotiation or policy planning processes. For many citizens, joining La Coordinadora was a response to the lack of democratic process in the policy negotiation.

This lack of citizen participation continued as Aguas del Tunari implemented the policy. The executives from Aguas del Tunari working in Cochabamba were foreign engineers with little understanding of local sentiments regarding privatization (Finnegan 2002). They were committed to implementing the plans for expanding the city's water system, and they clearly stated that those who could not pay the new price for water would be cut off from service. The message sent during the implementation of the policy was that public input and consultation were unnecessary and unimportant. Government and World Bank experts who understood the complexity of the market and the water technology were portrayed as the appropriate actors to determine solutions to water problems, while local expert opinions were disregarded.

During the implementation of the policy, consumers joined La Coordinadora mainly in response to the rise in water prices and the seizure of private water sources. La Coordinadora also included organized citizens with normative claims about the exclusionary and undemocratic nature of the implementation process. While the government gave rhetorical support to open dialogue, it dealt exclusively with legally accredited organizations, which many citizens felt were co-opted by the municipal government and did not represent their interests (Assies 2003). Therefore, La Coordinadora posed a challenge to the "instituted system of legally accredited representation, which failed to channel the concerns and interests of large sectors of the population" regarding the controversial privatization policy (Assies 2003, 32).

In response to exclusion from all aspects of the policy negotiation and implementation by Aguas del Tunari and the Bolivian government, La Coordinadora began to pursue its agenda outside of the official policy and legal system. La Coordinadora only achieved a place at the negotiating table after sustaining months of protest in the streets. When the government finally conceded to the demands of La Coordinadora, they agreed to create structures for ongoing inclusion of local community groups in future planning on water management. On the subject of the lack of consultation with local communities, Bolivian hydrologist and the UNESCO water expert Carlos Fernandez Jauregui asserts, "water legislation has to be based on consulting local people, as other laws are. If local culture, customs, and ways of life had been taken into account, all these problems could have been avoided" (cited in Slattery 2003).

While the Bolivian government refused to consider the needs and opinions of local organizations and citizens during the initial negotiation and implementation of the policy, the World Bank also disregarded the importance of communication with the public about the policy. This disregard is evident in the World Bank's analysis of the failure of water privatization in Cochabamba. According to one World Bank analysis of water privatization in Bolivia, the contract negotiated with Aguas del Tunari included a costly project that would bring water to Cochabamba from the Misicuni reservoir (World Bank 2002a). This project would be partly financed through price hikes and would eventually allow for improvements in service. Aguas del Tunari agreed to invest US$85 million in infrastructure, and had a goal to provide 24-hour service by the second year of the contract (World Bank 2002a). Since this contract was made behind closed doors with no public input and no information to the consumers, when the initial price hikes were implemented, citizens of Cochabamba believed they were simply being charged higher prices for the same subpar service.

One World Bank report attributes the failure of water privatization in Cochabamba to the public's misunderstanding of the intentions and plans of the company (World Bank 2002a). The report does not mention the lack of information available to the citizens about the contract, nor does it recommend using education about future privatizations to consumers as a possible remedy to the breakdown in communication during the implementation. The failure of the policy is blamed on public ignorance regarding the merits of private investment, and on local politicians who impeded the privatization process. The report refers to resistance by the mayor of Cochabamba to the price hikes as "political interference" by a local politician (World Bank 2002a, 2), not as a legitimate response to a policy that was unpopular, secretive, and potentially life-threatening to the citizens of Cochabamba. The World Bank report concludes that the main lesson to be drawn from the failure in Cochabamba is that consumers do not like to pay higher rates for the same subpar water service they experienced under the public system (World Bank 2002a). This analysis reveals the problematic nature of policy implementation that shuts out citizen participation and ignores the rights of citizens to information about policy designs.

The implementation of the policy relied on socially constructed notions about its target population (the water users in Cochabamba). The policy constructed the citizens of Cochabamba as consumers, not citizens who might take action about a policy that affected their daily lives. It also characterized them as likely to waste water. Promoters of privatization often claim that when consumers do not pay high prices for water, they will waste it (Schultz 2005a). This is reflected in the comments of former World Bank President, James Wolfensohn, "the biggest problem with water is the wastage of water through lack of charging" (Cited in Schultz 2005a).[5] In Cochabamba, where water is piped for two hours every other day into homes with no washing machines, dishwashers, lawn sprinklers or other water-intensive luxuries, this is a problematic claim to sustain empirically (Shultz 2005a). As a number of private water companies have discovered, many water users in Bolivia are more likely to conserve water than to waste it.

Following privatization of water in the Bolivian city of El Alto, profits from domestic water use fell short of the company's expectations. This was partly due to the large numbers of rural people who had moved to the city and were accustomed to conserving water (Forero 2005). Rate hikes also forced many residents to cut back on water use. The company was considering incorporating a public relations campaign to *promote* water use in El Alto to counteract the drop in consumption that resulted from the rise in tariffs and the cultural practice of conserving water (World Bank 2002a).[6] In this situation, citizens are constructed as ignorant consumers who do not understand the true value of water and who must be manipulated by pricing and public relations campaigns to use the "appropriate" amount of water in their homes. As seen in the implementation of the policy, the social construction of target populations carries implications for citizen participation.

Social Construction of Knowledge Within the World Bank's Water Policy Design

Socially constructed knowledge is embedded within the design of policies (Schneider and Ingram 1997). Water privatization policy involves socially constructed notions about water, government involvement in utilities, and the role of professional and expert communities in determining appropriate policy. The way water is viewed influences approaches to water-related policy. Water has evolved from being viewed as a common good to being seen as a commodity that is managed according to economic principles (Finger and Allouche 2002).[7] Water policy that treats water as a commodity assumes the market price for water is an inherently fair price that reflects the true value of water. This commodification of water is seen in World Bank privatization policies, which advocate a "full-cost recovery" pricing system requiring that consumers pay a price for water that reflects the cost of the water, the cost of infrastructure development, and the necessary profit for corporate investors. This conception of water is drawn from professional knowledge regarding economic efficiency as the most important concern for water management.

Some analysts question whether an approach to water policy that treats water solely as an economic good is compatible with long-term goals of responsible water management (Bauer 2004). Carl Bauer's in-depth research of Chilean water markets[8] raises questions about using conventional economic approaches to water. Bauer advocates a more qualitative and interdisciplinary approach to the "economics of water," which "emphasizes the institutional, legal, social, and political aspects of economic analysis" that are often ignored in conventional economic approaches to water policy (Bauer 2004, 3). Bauer asserts that conventional economic approaches, including neoliberal economics, are often limited to quantitative, technical methods. These approaches are rooted in orthodox neoclassical economics that do not incorporate historical or interdisciplinary analyses, such as institutional economics, political economy, or ecological economics. The result is a "narrow" conception of water, which ignores the legal and institutional arrangements that are necessary for successful, efficient, and equitable water resource management (Bauer 2004, 11).[9]

From this perspective, privatization is a simplistic solution to a complex problem that is not solely economic in nature. The growing scarcity of water worldwide and the problems associated with inadequate access to water and sanitation in poor countries are political problems that purely economic solutions are inadequate to address (Finger and Allouche 2002). The rationale that privatization alone is the most economically efficient manner to handle water and sanitation services ignores the nonmonetary costs of privatization, such as the costs to the welfare of the citizens, the loss of national sovereignty to multinational corporations and international financial institutions, and the damage done to democratic principles of accountability and transparency. Efficiency may not be the primary concern of water consumers, for whom issues of water quality, equitable access, and affordability are likely to be more relevant. Further, the value of water is not fully captured in scientific or economic studies alone, as values and ideas about quality of life standards vary. Water is not merely a product that is consumed; it is an essential element in public health, economic development, industry, and agriculture. Water has also historically been imbued with religious and cultural meaning that cannot be quantified (See Shiva 2002).

The debate at the second World Water Forum in The Hague in 2000[10] over whether to define water as a human "right" or a human "need" reveals how important common understandings are for water policy. Defining water as a human right implies obligations on the part of states and the international community to ensure that it is available to everyone. Defining water as a human need, as it was ultimately defined at the World Water Forum, justifies private control over the resource (Barlow and Clarke 2002). The decision to define water as a human need was decried by water activists and NGOs that support adopting an international definition of water as a social good, a human right, and a part of the global commons that is uniquely suited for public and local control.[11] In Bolivia, La Coordinadora promotes an alternative vision for local water management that views water as "a social good [and] a sensitive social issue" (De La Fuenta 2003). La Coordinadora favors local public control of the water system that is independent of the federal government and private foreign companies, and that incorporates "active participation in the administration and control of public and social services" (La Coordinadora 2005).

An important aspect of the rationale for privatization is the assertion that private companies, not governments, should carry out water delivery and sanitation services. In the rationale for privatization, governments are often portrayed as inefficient and inappropriate for managing this kind of service, while the private sector is described as efficient and therefore better suited for the task. According to the World Bank, providing water is a daunting task for poor nations; it is expensive, and corruption and bureaucracy frequently plague public utilities. In the case of Cochabamba, the city's public water system was not effective in delivering water to residents and was straining to meet demands for service expansion. Therefore, Cochabamba was particularly vulnerable to a conception of publicly owned utilities as inefficient.

Citizens and organized movements that resist water privatization often hope to bring the government back into service provision.[12] Many citizens involved in antiprivatization movements question whether private ownership is the solution to government inefficiency, particularly in countries where years of fiscal austerity measures have cut funding for social services like water and sanitation. There is also growing distrust of private foreign control over natural resources and services that are essential to life, as private control is viewed as less accountable and less open to citizen participation than publicly owned utilities. For those who view water as a fundamentally public resource, removing government from provision of water services is problematic despite concerns about government inefficiency.

Professionals and expert communities, such as the World Bank and multinational water companies, play a role in the social construction of water and of private ownership of utilities. The institutional culture of these communities and their agreed-on notions about the appropriate way to manage water directly affect policy designs. Policymaking is seen to be more objective when experts play a large role in the creation and implementation of the policy, and when utilitarian rationality is the dominant value that guides policy (Schneider and Ingram 1997). Through the use of the scientific method to determine the facts of any given policy situation, the power of social constructions is supposedly diminished, and solutions to social problems are discovered in an objective way. This process creates an illusion of neutrality and implies a transcendence of the pitfalls and inequalities commonly associated with policymaking. From this perspective, scientists, professionals, and academics emerge as the appropriate experts to be consulted in policymaking, while local citizen input and knowledge is often viewed as unnecessary.

Scientific and professional policy design is not value-neutral. It does not necessarily escape the pitfalls of degenerative politics. Scientific and professional expertise often relies on a particular type of knowledge that is limited to utility and rationality considerations. This approach to policy typically does not consider values and cultural factors that cannot be measured empirically (Schneider and Ingram 1997). Scientifically designed policies can serve interests that run counter to the public interest. They can reinforce unequal and unjust relationships.

Policies are inherently normative in nature in that they are constructed according to a set of normative understandings (Fischer 2000). Therefore, claims of expertise based on technical knowledge alone do not always mesh with the "real world" of public policy (Fischer 2000, 43). Just as science is not immune to social and political forces, neither are economic determinations of what constitutes good policy. Water privatization is accepted as sound policy largely based on expert analysis that considers market-rate pricing and private ownership to be superior to public control (World Bank 2005; Peet 2003). Policies based purely on a market rationale can overlook or ignore issues of equity, and they can delegitimize local knowledge and alternative policy elements.

The World Bank's water privatization policy in Bolivia provides an example of the implications of accepting scientific and professional constructions of knowledge, and of the ascendance of expert communities in public policy decision making. The tension between professional expertise and democratic governance is exacerbated when "policy communities" gain influence over issues that affect the public (Fischer 2000; Schneider and Ingram 1997). Policy communities form when experts come to agreement on the problems, solutions, and desired outcomes of policy situations. Within these communities, outside knowledge or perspectives are often considered to be irrelevant. Policy communities have reputations for being knowledgeable, prestigious, and best suited to formulate policy that deals with complex social problems. By essentially placing themselves between politicians and the public, policy communities can allow politicians to avoid blame or responsibility for policy outcomes. The policy community can insulate itself from scrutiny by emphasizing the technical aspects of policy that the general public does not understand.

The World Bank is a policy community that advances a coherent set of solutions to the pressing problems of developing countries. International organizations like the World Bank have become sources of authority in the international system, drawing power from the legitimacy of the "rational-legal authority" they embody and the control they have over technical expertise and information (Barnett and Finnemore 1999, 707). Bureaucracies like the World Bank appear to be depoliticized because they present themselves as impersonal, technocratic, and neutral. They derive authority from this image, and modern society views this type of rational-legal authority as legitimate (Barnett and Finnemore 1999). The World Bank also derives power from its ability to pressure governments into adopting their policy recommendations, as seen in Bolivia. Encouraging the empowerment of experts and professionals in policymaking can denigrate the role of the ordinary citizen in the policy process, as there is less need to educate and involve ordinary citizens when decision making and advising are perceived as primarily the role of epistemic communities.

Although social constructions of knowledge are powerful and can become entrenched, they are also dynamic and frequently contested. The response to water privatization in Cochabamba is an example of how different ideas about water, private control of services, and the role of experts can emerge and challenge the dominant paradigm for understanding these issues. The reversal of privatization in Cochabamba represented a major shift in policy. Large-scale policy change can occur when beliefs and values (the policy image) interact with the political institutions which are the venue for policy action (Baumgartner and Jones 1991). Policy shifts occur when many people change their views on an issue, thus changing the policy image. By reframing the image of an issue, groups can move the issue to a venue that is more favorable to their particular image of the issue (Mintrom 1999). In the case of Cochabamba, the community challenged the image of water as a commodity that should be subject to market prices. Instead they framed water as a basic human right that ought to be subject to local community control in a manner that ensures equity in water management. This reframing of the issue, combined with the massive pressure on the government by protesters, led to the canceling of the private contract.

Policies based on scientific and professional perspectives send messages to citizens about the unimportance of civic participation and local knowledge. These messages are one component of the overall policy design that includes goals, implementation structures, and socially constructed notions of target populations and knowledge. The social movement that emerged in Cochabamba effectively challenged the policy itself, the social constructions the policy reproduced, and the role of the World Bank in Bolivia.

Persisting Criticisms of the World Bank Reflected in the Cochabamba Privatization

Although protests and criticisms about water privatization in Cochabamba were mostly aimed at the Bolivian government and Aguas del Tunari, La Coordinadora members also called attention to the World Bank's role in pushing privatization despite widespread community opposition to the policy. They cited the lack of civil society participation in decision making about reforms to the water sector and the lack of ongoing consultation with civil society groups in the implementation of the policy as evidence of a lack of democratic participation in World Bank practice.

With the diffusion of democratic norms in the international arena, international institutions like the World Bank face increasing scrutiny regarding issues of transparency and accountability (Karns and Mingst 2004). The World Bank is a public institution that was designed to facilitate global collective action. The World Bank is also the key international organization in the area of development. Its role has expanded from financing infrastructure projects in developing countries to reforming entire states by promoting a specific brand of development (Finger and Allouche 2002).

Water projects have always been part of the World Bank's project portfolio, mainly in the form of large-scale dams and hydropower projects (Moore and Sklar 1998). The World Bank maintains an emphasis on large-scale capital-intensive water projects, despite a rhetorical commitment to shifting resources away from these endeavors and toward addressing the pressing need for projects that improve water supply and sanitation for the world's poor. The Bank's water supply and sewerage projects have a poor track record in terms of efficiency, improved service, and improved access to the poor (Moore and Sklar 1998).

An internal review carried out by the World Bank's Operations Evaluation Department (OED) surveyed 129 water supply and sewerage projects funded by the World Bank. Over half of the projects suffered from cost overruns, low economic rates of return, and excessive loss of water. Only two of the projects reviewed demonstrated success in improving conditions for poor households (Moore and Sklar 1998). Internal studies reveal that World Bank–funded water projects frequently benefit the middle and wealthier classes and large-scale farmers rather than the poor and small-scale farmers (Moore and Sklar 1998).

The national governments of borrowing countries share responsibility with the World Bank for some of these policy failures and for the lack of civil society involvement. Local NGO access to the World Bank and the influence of local grassroots groups is strongly influenced by the degree of democratization in the national politics of the borrowing country (Fox and Brown 1998). There is also the ongoing risk that borrowing governments will not implement World Bank policies in the appropriate manner. These concerns are particularly applicable to the role and actions of the Bolivian government in the Cochabamba privatization.

Conclusion

The policy design of water privatization in Bolivia relied on a social construction of knowledge that prioritized utilitarian concerns over issues of equity and democracy. The defective policy was negotiated and implemented in a deceptive and unjust manner, and it shifted control over a vital social service to a private company with little accountability to the citizens it served. As a result of citizen exclusion from the policy design and implementation process, a movement of concerned citizens emerged, which successfully reversed the privatization and returned the water system to public control. The response from the people of Cochabamba to water privatization is indicative of the power of social movements when faced with policies that are inequitable and undemocratic. By reversing the privatization of water and demanding self-governance of local water systems, the citizens of Cochabamba challenged more than a specific policy outcome—they challenged a policy design that undermined their right to democratic participation and active citizenship regarding an issue that is essential to life and development. They also posed a challenge to the notion of economic efficiency as the appropriate guiding principle in water policymaking, as concerns about efficiency often ignore issues of equity in water access.

La Coordinadora represented a new form of social movement organizing in Bolivia that was based on territorial concerns and resources rather than on union affiliation (Garcia Linera 2004). The battle over water privatization policy in Cochabamba opened up a larger discussion in Bolivia over control of natural resources and other public goods, state accountability for services, and issues of democracy and representation. Similar protests against water privatization in the Bolivian cities of El Alto and La Paz and the national "gas wars" over the question of foreign ownership of the country's natural gas reserves point to a growing antiprivatization movement in Bolivia. In May 2006, recently elected president Evo Morales nationalized the nation's oil and gas reserves in response to growing public disapproval of foreign ownership and control of natural resources.

With the cancelation of the contract with Aguas del Tunari, control of the Cochabamba water system reverted back to the local community and the municipal government. However, removing the private company from Cochabamba has not solved the problem of inadequate access to safe water in the city and surrounding neighborhoods. The new body charged with managing the water system has struggled with bad infrastructure and service demands that are difficult if not impossible to meet without major investment of resources. Such major investment from the public or private sectors of the international financial community is not likely to be forthcoming in the foreseeable future, nor does Bolivia itself have the requisite financial resources. The public water utility in Cochabamba enjoys support from the population and is making progress toward sustainable and socially responsible management of water resources (Assies 2003). However, one analyst describes the victory of La Coordinadora as "an illusion," citing the fact that many of the city's poorest residents continue to get their water from dirty pools and from private vendors of questionable integrity (Laserna 2005).

Despite these criticisms, La Coordinadora clearly succeeded in drawing international attention to issues of water rights, the role of multinational corporations in developing countries, and the politics of privatization. This movement showed the Bolivian government, the World Bank, and the international community the potential consequences of shutting citizens out of policy decisions that impact the daily lives of people who are already living in a precarious condition of poverty and underdevelopment. The recent "gas wars" reveal an ongoing desire for a form of political participation in Bolivia that encompasses more than traditional electoral politics; the citizens of Bolivia are repeatedly demanding a voice in the management of the national patrimony. In this way, the battle over water in Cochabamba was a precursor to a continued struggle in Bolivia over who will own and obtain the benefits from natural resources, who will provide services that are essential to the survival and well-being of society, and most importantly, who has the right to decide these matters for the entire nation.

Notes

1. See Moore and Sklar 1998 for data on the performance of water projects financed by the World Bank.

2. Bechtel owns 50 percent of International Water; International Water owns 55 percent of Aguas del Tunari.

3. See http://democracyctr.org/bechtel/waterbills/index.htm for copies of actual water bills from Aguas del Tunari showing these increases.

4. See Assies 2003 for the history of La Coordinadora.

5. Cited in Schultz 2005a, from comments made by Wolfensohn to a Washington, DC news conference, April 12, 2000.

6. The company had asked state regulators for permission to raise monthly fees again. This request was rejected, but they did receive permission to increase the new connection fee for water and sewage to US$450, a fee that was far too expensive for most citizens to pay (Braun 2005; Forero 2005).

7. See Bauer 2004 for a detailed history of this shift to defining water as an "economic good."

8. Water policy in Chile has been the world's leading example of the free-market approach to water management.

9. Bauer criticizes the World Bank's unwavering support for water privatization as rooted in ideological justifications, not empirical evidence.

10. The World Water Forum is an annual meeting of lobby organizations, such as the Global Water Partnership, the World Bank, multinational water corporations, and other representatives from the global water sector.

11. See, for example: The Blue Planet Project of the Council of Canadians (http://www.blueplanetproject.net/english/links/), and Public Citizen's Water For All Campaign (http://www.citizen.org/cmep/Water/).

12. See Shiva 2002 for movements in India that resist both corporate and government control of water in favor of local, community-managed systems.

References

Assies, Willem. 2003. "David versus Goliath in Cochabamba: Water Rights, Neoliberalism, and the Revival of Social Protest in Bolivia." *Latin American Perspectives* 30, no. 3:14–36.

Barlow, Maude, and Tony Clarke. 2002. *Blue Gold.* New York: The New Press.

Barnett, Michael N., and Martha Finnemore. 1999. "The Politics and Pathologies of International Organizations." *International Organization* 53, no. 4:699–732.

Bauer, Carl J. 2004. *Siren Song: Chilean Water Law as a Model for International Reform.* Washington, DC: Resources for the Future.

Baumgartner, Frank, and Bryan Jones. 1991. "Agenda Dynamics and Policy Subsystems." *Journal of Politics* 53, no. 4:1044–1074.

Bennett, Vivienne, Sonia Davila-Poblete, and Nieves Rico, eds. 2005. *Opposing Currents: The Politics of Water and Gender in Latin America.* Pittsburgh, PA: University of Pittsburgh Press.

Braun, Will. 2005. "Bolivians Win Water War II." *Znet,* January 16. www.zmag .org/content/print_article.cfm?itemID=7033&sectionID=52.

Chong, Alberto, and Florencio Lopez-de-Salines. 2004. "Privatization in Latin America: What Does the Evidence Say?" *Economia,* Spring, 37–111.

De la Fuente, Manuel. 2003. "Presentation on the Struggle in Cochabamba, World Social Forum 2003." http://www.publiccitizen.org/cmep/Water/cmep_Water/ reports/bolivia/articles.cfm?ID=10608.

El Diario (La Paz, Bolivia). 1997. "Organismos Multilaterales Presionan al Gobierno: Condonaran $US 600 milliones de Deuda si Privatizan SEMAPA de Cochabamba." July 1, 5.

Finger, Matthias, and Jeremy Allouche. 2002. *Water Privatization: Trans-National Corporations and the Re-Regulation of the Water Industry.* London: Spon Press.

Finnegan, William. 2002. "Leasing the Rain," *New Yorker,* April 8.

Fischer, Frank. 2000. *Citizens, Experts, and the Environment: The Politics of Local Knowledge.* Durham, NC: Duke University Press.

Forero, Juan. 2005. "Latin America Fails to Deliver on Basic Needs." *New York Times,* February 22.

Fox, Jonathan A., and L. David Brown, eds. 1998. *The Struggle for Accountability: The World Bank, NGOs, and Grassroots Movements.* Cambridge, MA: MIT Press.

Galiani, Sebastian, Paul Gertler, and Ernesto Schargrodsky. 2005. "Water for Life: The Impact of Privatization of Water Services on Child Mortality." *Journal of Political Economy* 113, no. 1:83–119.

Garcia Linera, Alvaro. 2004. "The 'Multitude.'" In Oscar Olivera and Tom Lewis, eds., *Cochabamba! Water War in Bolivia.* Cambridge, MA: South End Press, pp. 65–86.

Gill, Lesley. 2000. *Teetering on the Rim: Global Restructuring, Daily Life, and the Armed Retreat of the Bolivian State.* New York: Columbia University Press.

Karns, Margaret P., and Karen A. Mingst. 2004. *International Organizations: The Politics and Processes of Global Governance.* Boulder: Lynne Rienner Publishers.

Klein, Michael. 2003. "Where Do We Stand Today with Private Infrastructure?" *Development Outreach: Special Report: The Role of the Private Sector.* The World Bank Institute. www1.worldbank.org/devoutreach/march03/article.asp ?id=190.

La Coordinadora. 1999. "Texts of the Coordinadora del Agua of Cochabamba." http://www.nadir.org/nadir/initiativ/agp/cocha/agua.html.

La Coordinadora. 2005. "Open Letter to the World Bank, Inter-American Development Bank and the German Cooperation GTZ." March 6, 2005. www.citizen .org/print_article/cfm?ID=13065.

Laserna, Roberto. 2000. "Cochabamba: La Guerra Contra el Agua." *Revista del Observatoria Social de America Latina* 2:15–20.

Laserna, Roberto. 2005. "Etica del Agua," *Los Tiempos,* May 22. www.geocities .com/laserna_r/eticagua.html.

McDonnell, Patrick J. 2006. "Bolivian Leader Nationalizes Fuel Industry," *Los Angeles Times,* May 2, A21.

Meyer, David. 2005. "Social Movements and Public Policy: Eggs, Chickens, and Theory." In David S. Meyer, Valerie Jenness, and Helen Ingram, eds., *Routing Opposition: Social Movements, Public Policy, and Democracy,* Minneapolis: University of Minnesota Press, pp. 1–38.

Mintrom, Michael. 1999. *Policy Entrepreneurs and School Choice.* Washington, DC: Georgetown University Press.

Moore, Deborah, and Leonard Sklar. 1998. "Reforming the World Bank's Lending for Water: The Process and Outcome of Developing a Water Resources Management Policy." In Jonathan A. Fox, and L. David Brown, eds., *The Struggle for Accountability: The World Bank, NGOs, and Grassroots Movements,* Cambridge, MA: MIT Press, pp. 345–390.

Olivera, Oscar. 2004. *Cochabamba! Water War in Bolivia.* Cambridge, MA: South End Press.

Painter, James. 2005. "Why is Bolivia in Turmoil?" *BBC News,* June 3. http:// news.bbc.co.uk/2/hi/americas/4604173.stm.

Peet, John. 2003. "Survey: Priceless," *Economist* 368, issue 8333: 3.

Primera Plana (La Paz, Bolivia). 1996. "Banco Mundial Es Claro: Sin Privatización de SEMAPA, No Hay Agua Potable para Cochabamba." February 29, 10.

Public Citizen. 2003. "Water Privatization Case Study: Cochabamba, Bolivia." *Public Citizen,* Washington, DC. http://www.citizen.org/documents/Bolivia _(PDF).PDF.

Schneider, Anne, and Helen Ingram. 1997. *Policy Design for Democracy.* Lawrence, Kansas: University Press of Kansas.

Shiva, Vandana. 2002. *Water Wars: Privatization, Pollution, and Profit.* Cambridge, MA: South End Press.

Shultz, Jim. 2003a. "Bolivia: The Water War Widens." *NACLA Report on the Americas* 36, no.3:34–37.

Shultz, Jim. 2003b. "Bolivia's War over Water." www.democracyctr.org/bechtel/ the_water_war.htm.

Shultz, Jim. 2005a. "The Right to Water. Fulfilling the Promise," forthcoming in Economic, Social and Cultural Rights in Latin America: From Theory to Practice. http://www.democracyctr.org/bechtel/righttowater.htm.

Shultz, Jim. 2005b. "The Politics of Water in Bolivia," *The Nation.* www.thenation .com/doc/20050214/shultz.

Slattery, Kathleen. 2003. "What Went Wrong? Lessons from Cochabamba, Manila, Buenos Aires, and Atlanta." *Annual Privatization Report 2003, Reason Public Policy Institute.* www.rppi.org/apr2003/whatwentwrong.html.

Stiglitz, Joseph E. 1999. "The World Bank at the Millennium," *Economic Journal* 109, no. 459:F577–F597.

Stiglitz, Joseph E. 2002. *Globalization and Its Discontents.* New York: W.W. Norton and Company.

Strange, Susan. 1996. *The Retreat of the State: The Diffusion of Power in the World Economy.* Cambridge, UK: Cambridge University Press.

Tarrow, Sidney. 1998. *Power in Movement: Social Movements and Contentious Politics, Second Edition.* Cambridge, UK: Cambridge University Press.

Udall, Lori. 1998. "The World Bank and Public Accountability: Has Anything Changed?" In Jonathan A. Fox and L. David Brown, eds., *The Struggle for Accountability: The World Bank, NGOs, and Grassroots Movements,* Cambridge, MA: MIT Press, pp. 339–436.

World Bank. 1993. "Water Resources Management: A World Bank Policy Paper." Washington, DC: World Bank.

World Bank. 1999. "Bolivia Public Expenditure Review." Washington, DC: World Bank. June 14.

World Bank. 2002a. "Bolivia Water Management: A Tale of Three Cities." *Precis.* Washington, DC: World Bank Operations Evaluation Department.

World Bank. 2002b. "World Bank Brief: Water Supply and Sanitation." http:// www.worldbank.org/watsan/WB percent20Brief.htm.

World Bank. 2004a. "The World Bank Group's Program for Water Supply and Sanitation." *Water Supply and Sanitation Sector Board.* The World Bank Group.

World Bank. 2004b. "Millennium Development Goals." http://ddp-ext.worldbank.org/ext/MDG/home.do.

World Bank. 2005. "Privatizing Water and Sanitation Services—Papers and Links—Privatization, Infrastructure, and Business Environment." http://rru.worldbank.org/Themes/Privatization/.

World Health Organization. 2004. "Water, Sanitation and Hygiene Links to Health." http://www.who.int/water_sanitation_health/publications/facts2004/en/index.htm.

8

Modernizing Mountain Water: State, Industry, and Territory

Ismael Vaccaro

The Valley of Lillet is part of the upper watershed of the Llobregat River. It is located in the northern corner of the province of Barcelona at the first ramparts of the central Catalan Pyrenees. The river has always been at the geographical and social center of the valley. Since the Middle Ages, its canyons have constituted the best path to the lowlands markets, and since the advent of industrialization, its waters have powered the plethora of factories that emerged along its course.

The river begins as a small stream in the Pyrenees. It becomes the backbone of the region as it flows toward the lowlands. This anatomical analogy is used not only because of the river's geographical location and significance, but also because it has historically been a social artery, carrier of people, ideas, and wealth.

In the last two centuries the rugged territory of the Pyrenees has undergone numerous and significant changes. In the Valley of Lillet these rearrangements of the social and the ecological landscape translated into the weakening of old networks of isolated farms and range villages, as factories and mines attracted the population to the valleys. Industrial infrastructures were abandoned after the oil crises of the 1970s. At the end of the twentieth century the depopulated ranges were overtaken by protected areas and ski resorts.

The Pyrenean inland thus presents an excellent opportunity to study the transition from tradition to modernity (from subsistence to mass production) and from modernity to postindustriality (from mass production to services and leisure). This transformation offers a privileged ground to discuss the process of intellectual reconstruction and physical appropriation of the water resources of the rural areas of Spain. In the framework of this book, devoted to the concept of "equity," this chapter offers an example that characterizes a significant number of processes of national

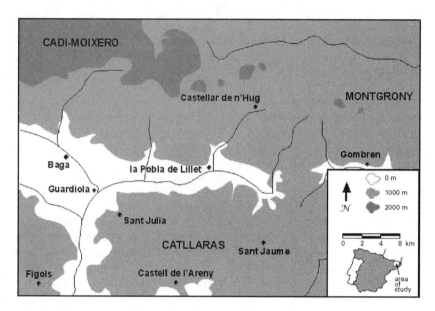

Figure 8.1
The Valley of Lillet and its surroundings

modernization that have not paid much attention to environmental justice and social equity. Modern nations are conceived, governed, and managed from their urban centers. The political and economic power of these centers has often resulted in the subjugation of population and resources from their rural areas to the "superior" needs of a national community that is mostly urban.

In this chapter I will briefly summarize this process of social transformation in order to provide local political and economical context for the analysis of water policies.[1] The main goal, however, is to discuss the role that water, directly or indirectly, played in these transformations.

As in most developed countries undergoing the creation, expansion, and consolidation of the modern state, in the last two hundred years in territories within Spain natural resources experienced a succession of redefinitions. These reconceptualizations affected the way in which territory and resources were perceived and the way they were managed. The resulting policies and the ideological backgrounds that sustained them translated into legislative initiatives, appropriation schemes and development projects. These processes inextricably linked resources and people to the concept of national territory and sovereignty. This progression has been called territorialization (Vandergeest and Peluso 1995; Brenner 1997; Hannah 2000; Braun 2002).

These territorialization initiatives and projects were mostly designed, financed, and implemented by representatives of the mainstream national society. Modern mainstream national societies are articulated by and around political decision centers, productive poles, and demographic concentrations located in urban areas. The global landscape has gradually become a network of individuals and infrastructures that facilitate the transmission of information, culture, and values (Castells 1996; Nel.lo 2001; Agrawal and Sivaramakrishnan 2003). Footprints of cities expand well beyond their immediate territory. Because they are hardly self-sufficient, they are forced to collect the resources they need in their extensive areas of influence (Williams 1972; Cronon 1991; Gandy 2002). Water is chief among these resource needs. In all modern urban societies, water management quickly became a national priority. Urbanely conceived and controlled water management programs, however, seldom take into account the needs of rural populations.

The objective here is to approach the changes in water management that occurred in Spain during the nineteenth and twentieth centuries and, most importantly, to identify the consequences of these changes at a local level: what the modern culture of water represented for the inhabitants of the Valley of Lillet.

Economic Transformations

At least since the Middle Ages these valleys have been densely populated, and there has always been a tension between dispersed and concentrated patterns of habitation. In other words, these areas presented two alternative ways of life: larger villages mainly located at the bottom of the valleys versus the networks of isolated farms, hamlets and very small villages dispersed across the ranges.

As in many other mountainous landscapes all over the world, these rural networks resulted in an intensive use and management of all ecological levels of the range (Murra 1975; Netting 1981; Saberwal 1999; Bauer 2004). Obvious anthropogenic features such as ruins, paths, agricultural terraces, and fences are scattered all over the landscape. Not so obvious anthropogenic ecosystem effects such as altered forest species composition or human-maintained pastures are also a fundamental part of these mountains.

The demographic record demonstrates that the precarious equilibrium between valley villages and farms and small villages from the ranges began to change at the end of the nineteenth century. During the nineteenth century most of the small range villages disappeared. By the 1950s the percentage of population living on farms had declined to an insignificant 5 percent. This percentage plunged further during the next decades to reach the contemporary level of 2 percent. In earlier times, in the nineteenth century, approximately 20 percent of the population had lived on isolated farms.

This period of demographic decay in the ranges does not match a similar trend in the valleys. At the end of the nineteenth century, but especially during the first half of the twentieth, coal mines and several types of factories mushroomed all over the area (textile and concrete). The factories were drawn to these mountains because of the area's potential for inexpensive hydropower production, raw materials, and labor force. The labor demand of this fast-growing industrial complex soon exceeded the demographic potential of these mountains. During those years, the villages, almost towns, in the valleys absorbed the population flooding down from the farms and also drew immigrants from all over Spain. The 1950s and 1960s were the golden years of la Pobla de Lillet, the main town of the valley.

The oil crises of the early 1970s represented the end of the industrial era in the Catalan mountains. Even after adding in the additional transportation costs, coal could be obtained more cheaply from the open mines of South Africa. Textile factories could not compete with low-cost Moroccan products. The concrete factory moved its activities to southern Spain where a new state-of-the-art facility had been built, which was well- connected to the international trade networks. Globalization essentially sucked the industrial life out of these valleys. The lack of easy interconnectivity with a globalized world economically condemned mountain industries. This industrial collapse translated into an acute process of depopulation in the towns.

The last quarter of the twentieth century resulted in an abandoned landscape. The villages had lost more than 50 percent of their populations. Most of the farms were empty. For the most part, only residents beyond their productive years remained in the valley. The absence of human activities resulted in an unprecedented expansion of the forest and the natural or induced return of rare or locally extinct animal species such as red deer, chamois, and roe deer.[2] After the 1970s, large areas of the Pyrenees were declared conservation areas through conservation policies that conceptualized this territory as undisturbed and deserving of state protection from degradation. Ironically, these new nature preserves had literally grown up on top of the remains of two abandoned ways of life (agrarian and industrial).

What was the role of water in such a process? We next highlight the ways in which water policies have affected, directly or indirectly, literally or ideologically, the Valley of Lillet. As we will see, these consequences transformed the economic organization, the social structure, and the ecology of these mountains. Every socioeconomic period can be associated with a specific approach to water management and to a particular type of water culture.

Water Policies in Spain

In order to understand the transformations that water management experienced in Spain during the process of modernization of the country, we need, first, to understand the evolution of the state's water policies. An analysis of the legal framework and the ideological background that informed these policies is a fundamental step to understanding the successive shifts that water in Spanish culture has experienced in the last two centuries. Law and policies are basic tools of social engineering and deserve careful analytical attention. They carry cultural assumptions and perpetuate and create power differentials (Shore and Wright 1997; Abramson and Theodossopoulos 2000; Wickham and Pavlich 2001).

A thorough examination of the national perspective on water management provides the broader political context for a local analysis. This analysis makes possible an examination of the social and ecological consequences of such policies. In modern societies, water policies have been implemented through four main managerial fields: irrigation, power production, human consumption, and environmental conservation. Since their consolidation as political monopolies, modern nation-states have produced public policies for each of these fields. Obviously, their chronology is completely diverse. In Spain, for many decades, water policies have been a national priority. This, however, was not always the case.

In the Iberian Peninsula, more than most European countries, water availability has become an obsession. Specific topographic and climatic conditions in Spain make water a scarce commodity. Desertification is both a discursive and a tangible possibility. As Portabella writes in his contribution to Arrojo's edited book (2004):

Today water animates and divides the different communities that constitute this country of ours . . . Deserts grow, contamination of our rivers is increasing, and a foolish speculation, especially at the shores of the Mediterranean coastlines, are making this issue detonate in the street, in the classrooms, amongst the political parties, and institutions of all sorts. (Arrojo 2004, 15)[3]

In Spain, as in other areas of the world, water control is an essential issue that has consequences for fundamental sectors of the country's economy and social life. These consequences do not respect administrative lines, national or international borders, or ecological zones (Hundley 1975, 1992; Worster 1985; Blatter and Ingram 2001). The tensions emerging from the competition over water spread out beyond the field of policies and expand into environmental justice issues (Brown and Ingram 1987; Donahue and Johnston 1998; Hicks and Peña 2003).

A quick summary of the development of Spanish water law will give an idea of the attention devoted by the state to water issues, the instruments devised, and the goals pursued. It is through this legal chronology that I intend to identify the cultural and ideological changes that have occurred in the public culture of water throughout the last 150 years (Ortega 1979; Swyngedouw 1999, 2001; Sauri and del Moral 1999, 2001; Pérez 2004).

• Prior to 1860: Absence of public involvement
• 1860–1890: Early legislative attempts by the liberal state
• 1890–1911: Water reaches a central role as a cultural and political subject
• 1911–1932: Institutionalization of management. First systematic projects
• 1932–1936: Implementation and expansion with social emphasis
• 1939–1980: Irrigation, hydropower and development. Autocracy
• 1980–2004: New uses, new priorities, and old conflicts. Democracy and supranational legislation

Before the 1860s, the Spanish state, as a managerial entity, did not pay much attention to water. Its uses and management were left to private initiatives. After the 1860s, water issues started timidly to make their way into the public political arena. The focus was almost exclusively agrarian. However, as Spain was a liberal state that mostly relied on private property and private initiative, the laws promulgated were mostly offering incentives to build irrigation complexes. As the margin of profitability of such projects was very low, they did not attract much private investment.[4]

The twenty-year period between the 1890s and 1911 was characterized by a major shift in the role of water in mainstream Spanish politics. The "regenerationist" movement, a mixture of politicians, ideologists, and folklorists, turned water into a national priority. A country in social distress needed a path to regeneration and modernization.[5] The "regenerationists" movement and its main figure, Joaquin Costa, considered that the harnessing of water power could be the catalyst of such a transformation (Costa 1975). The renovation was also meant to imply a restructuring of the traditional property regimes and policies regarding natural resources management which had been previously marked by deep inequalities. The agrarian economy was supposed to shift from cereal production to more exportable Mediterranean horticulture (vineyards, vegetables, and fruit trees) that was more dependent on irrigation. The logical consequence of the importance attributed to the water question by this intellectual movement was to increase the involvement of the state in water management. In other words, water, in the mind of the regenerationists, had to be centralized and managed by the state in order to be modernized and used efficiently. For years, however, this shift occurred only in the discursive domain.

The National Plan of Hydraulic Uses, first approved in 1902 and revised in 1909, 1916, and 1919, was the first comprehensive and ambitious attempt to reformulate water issues as a national issue and a priority. It was never fully implemented, and its tangible effects were minimal. In 1911 a new Law of Large Irrigations Projects, although moderate in its formulation, became the first legislation to be really effective in expanding state control over water (Ortega 1979). This law officially states the jurisdiction and need for state intervention in water issues, and very importantly, affirms for the first time the right of the state in the general interest to expropriate land for water management purposes.

The period between 1911 and 1932 experienced little movement in national water policies. Although legislation on water management was repeatedly produced, the tangible results were minimal. A fundamental development of this period was the creation of hydrographic confederations.[6] Although these institutions did not become operative for a long while, they represented the first juridical manifestation of change. The geographical range considered by this legislation shifted from the local to the regional level. Therefore, this period witnessed the institutionalization of water management beyond the local level.

From 1932 until the late 1970s, successive governments attempted to implement all sorts of development regimes for irrigation and hydropower generation. The pre–Civil War governments emphasized land reform and resource redistribution,[7] but they also settled the modernizing characteristics that marked water policies until the consolidation of democracy. Across this forty-year period, however, the priorities gradually shifted from agriculture to power generation and industry. In terms of size and quantity, the majority of Spanish dams were built between the 1930s and the 1980s (Swyngedouw 1999). The hydrographic confederations played a fundamental role in watershed management.

The consolidation of hydropower changed the Spanish approach to large water infrastructure. The production of energy had a significantly larger potential for economic benefits than irrigation projects. This potential channeled significant amounts of investment from public and private sectors into dams and hydroelectric industries.

Franco's fascist regime also increased the centralization of water management. In this regard, the water transfer system between the rivers Tajo and the Segura confirmed the nationalization of water management. After this transfer, watersheds were not the only managerial units. Rivers and watersheds became a unitary national network. Water was now transferred from areas with water surplus to areas with water deficit (Pérez 2004).

The urbanization of Spanish society and the centralization of decision making resulted in a *de facto* urban appropriation of the water of the Pyrenean valleys. Arqué, Garcia, and Mateu (1982) and Maluquer de Motes (1998) describe the high concentration of hydropower plants in the narrow gorges of the Pyrenees. Production and management were visibly oriented to industrial and urban needs. In the name of the national interest and modernization, entire valleys of the Pyrenees were transformed by the Spanish state and hydroelectric companies into hydropower reservoirs (Arqué, Garcia, and Mateu 1982; Tarraubella 1990; Boneta 2003).

In Spain, in opposition to actions taken in some other countries,[8] the modern culture of water was connected to large infrastructure projects and provided with public funding until the beginning of the twenty-first century (Arrojo 2004). Even though the pace of dam construction has slowed considerably since the late eighties, the projects and the political movements have not lost this publicly funded and modernizing emphasis. The 1985 Law of Waters, while actualizing the 1879 law, fully incorporated important tenants of the regenerationist ideals. Due to their costs and infrastructural needs, state involvement and control is almost a necessity for large dams and canal construction.

The end of the seventies and the early eighties witnessed the imposition of new values and new uses for water. Potability had been on the agenda for a long while but was joined by issues such as environmental protection and leisure uses for water such as golf courses and ski resorts. This process of diversification coincided with a decentralization of some levels of water management from the national government to the regional governments.[9]

The end of the twentieth century brought an old debate to Spain's main political stage. The conservative government of the Partido Popular designed a new National Hydrologic Plan based on a massive implementation of the modernist paradigm: the construction of dozens of new dams and the implementation of interregional water transfers. These policies encountered ferocious local opposition and divided opinions from the regional governments.[10] This controversial plan, with considerable dependence on European Union funds, divided Spanish society for several years (Sauri and del Moral 2001). The 2004 electoral defeat of the Partido Popular at the hands of the Partido Socialista returned the National Irrigation Plan to the design stage.

At the same time, another entity joined the Spanish political and legal waterscape. The European Union started to legislate on water issues. The Frame Directive of Waters, approved in 2000, established the overarching guidelines that should regulate water management all over the European Union. Interestingly enough, the EU priorities are connected to preventing contamination, rehabilitating water, and to balancing demand and supply in order to foster rational management (La Calle 2004). The connection of the Water Directive with the Birds Directive and the Habitats Directive (1992) promotes ecosystem integrity and emphasizes this particular focus on environmental protection. Supranational legislation is fostering the shift of water culture from development to sustainability. The values of modernity are progressively tempered by postmaterialist values solidly in place in other countries of Europe. The European Union initiative introduces equity issues into Spanish water policies.

The coexistence of different productive goals and ideological frameworks, irrigation, power generation, non-utilitarian values, and so on, was an interesting consequence of the historical sequence of water policies. The need to irrigate, or the emphasis on hydropower production, did not disappear with the emergence and consolidation of conservation and tourism values. All of these different approaches to water management reflect existing needs of local, regional, and national economies. The challenge here is to articulate and integrate dissimilar needs. The evolution of institutional management in Spain during the last thirty years is a partial consequence of the need to articulate and reconcile these contradictions in practice.

Local Consequences of the National Political Culture of Water

Accounts from the nineteenth century describe the Valley of Lillet as a valley with a healthy network of waterways. Although densely populated, the quality of its waters was high.[11] Besides a few low-impact artisan mills using traditional technology, water was mainly used for agricultural and drinking purposes.

There were also a few irrigation systems in the area: small schemes carved by groups of individuals working together under the framework of a communal institution.[12] However, this mountainous area is not suitable for large-scale agriculture. The only type of agriculture traditionally encountered in the valley was subsistence agriculture with low levels of trade. This status quo offered an integrated landscape in which water and its uses were part of a local productive and ecological system. This productive system can be defined as traditional, characterized by ranching, subsistence agriculture, and domestic industries. Water was managed according to local needs. The functional integration of local ecosystems and local social systems with communal institutions often results in localities with a strong sense of cultural, ecological and productive identity (Rivera 1999; Orlove 2002; Strang 2004). In these areas, knowledge and identity are specific and locally contextualized (Castells 1997; Strang 1997; Peña 1998; González 2001).

This situation remained stable in the valley until the last twenty years of the nineteenth century when three of the largest textile workshops of the town mechanized. In other words, the industrial age had arrived in la Pobla de Lillet. In 1904, a massive concrete factory started its operations upstream. Ten kilometers downstream a massive coal mining complex was installed. The mines and the adjacent colonies[13] required significant amounts of water for mineral cleansing, energy production, and human consumption. Each of these industrial ventures developed its own infrastructure for hydropower generation: usually implying a network of water deviations through small dams or channels.

In these years, the economic emphasis of la Pobla de Lillet and most of the area shifted to mass production to cover the needs of the expanding cities of the lowlands and an export economy. The migration fluxes, the local consumption patterns, and the steady importation of goods solidly connected the valley to the national social and economic systems. The local economy entered the age of modernity.

These transformations happened at the same time that the population concentrated in villages was experiencing a significant increase. The amount and types of waste produced also increased dramatically, eventually affecting and degrading the riverine ecosystems.

The turn from the nineteenth century to the twentieth thus represented a shift in the managerial and consumptive patterns applied to water use. Significant amounts of the river's flow were being diverted to cover the energy needs of the factories. At the same time, the river was receiving a large amount of detritus. Because of this, its ecological integrity quickly became compromised. The consequences of the discussions about modernization, generalized at a national level, physically materialized across the Llobregat watershed. The Llobregat at the end of the nineteenth century was one of the most industrialized rivers in Europe. Dozens of manufacturers were flourishing on its banks. Each of these factories, mostly textile, depended on the river as its power source. As a result, the river was littered in its low- and mid-course with small, privately owned dams.

During the seventies and early eighties, three dams, large by Catalan standards, were built across the Llobregat watershed. Only fifteen kilometers downstream from la Pobla de Lillet, the Baells dam was constructed, and the San Ponç and La Llosa del Cavall dams quickly followed. The Mediterranean climate is not especially humid, but it has concentrated rainy seasons in spring and fall. These specific conditions have fostered dam construction as a way to control water influx and to take maximum advantage of its productive potential for agricultural purposes. However, the main economic incentive is still hydropower generation. These dams are part of the Catalan system of water-flow control.[14]

In summary, the river became a fundamental element of the regional economy. Industries, lowlands' agricultural fields, and urban dwellers became dependent on its waters and its power. The systematic character of water ecology invited an integrated management. The river was incorporated into regional networks far greater that any of the particular localities along its course. Jurisdiction and, consequently, the decision-making control over water management shifted from the local to the national arena. This integration, however, did not happen until a central power had become hegemonic across the whole country. Isolated economic enterprises did not have enough control and jurisdiction over land, resources, and people.

Until now we have taken a look at the direct local consequences of the transformations of the culture and the economy of water in the Valley of Lillet. As mentioned earlier, although we cannot discuss grand irrigation schemes or dams, a few small infrastructures were present in the valley: mainly water diversion schemes to fuel factory power generators. The Valley of Lillet, not being one of the areas potentially affected by large irrigation schemes, and not having large numbers of landless peasants, was not directly affected by the water policy discussions of the late nineteenth and early twentieth century. To understand the consequences of water policies on the management of the territory, we cannot reduce the scope of our analysis to the lands adjacent to, or directly affected by dams or irrigation schemes. Such a methodological approach would minimize the significant effects that state policies had in relatively remote areas of the Spanish landscape.

In accord with the "Regeneracionista" ideology, water policies were part of a master plan to revitalize Spain as a society and as a territory. This included a massive reforestation effort to control erosion. These reforestation programs were preceded by widespread campaigns of territorial public appropriation through expropriation or forced acquisition of land from private or communal ownership to public ownership. The building of numerous dams downstream and the regional dependence on them legitimated the expropriation of enormous extensions of land in the upper watersheds. Water policies became tools of national territorialization at several levels, implying not only a reorganization of the water uses, but also of the tools that helped to reconfigure the territory through the alteration of property regimes already in place. Hydrology, forestry, and geology as academic and corporate disciplines were integrated into the managerial and decision-making apparatus of the state (Bauer 1980; Gómez 1992; Casals 1996).

In the municipalities of the Valley of Lillet the expropriation fever affected large areas of the Cadi Range, the northern side of the valley. The southern side, for historical reasons, already belonged to the state and demanded no intervention (Vaccaro 2005). The Forest Patrimony of the State, and specifically its Forest Hydrologic Service, managed these lands for decades until the creation of the National Institute for the Conservation of Nature (ICONA)[15]. The ICONA was in charge of the expropriated

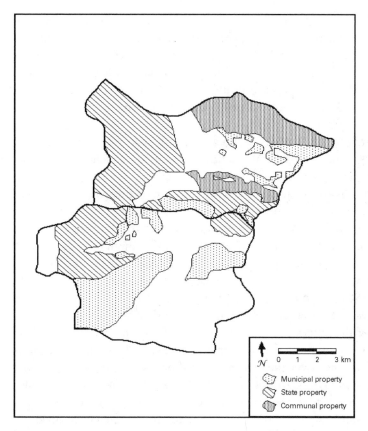

Figure 8.2
Collective property regimes in the municipalities of la Pobla de Lillet (south) and Castellar de n'Hug (north).

area until its transfer to the Generalitat de Catalunya, the autonomous regional government (Bauer 1980; Gómez 1992).

Diagonal stripes in figure 8.2 mark the areas seized by the state between 1900 and 1960. The reasons for such acquisitions and expropriations were essentially the same throughout this extended period of time: to protect the watershed from erosion and foster scientifically regulated reforestation. Interestingly enough, forest engineers were in charge of implementing such protective policies. The forest department had become entrenched with water policies by virtue of having the responsibility of protecting the watersheds from erosion (Mangas 1984; Gómez 1992).

One of the main tasks of the corps of engineers was to avoid the clogging of river streams and dams with debris from the deforested slopes of the Pyrenees. The rationale was that traditional practices were harmful to environmental integrity and naturally conducive to deforestation and erosion. Only top-down, scientifically oriented control was viewed as adequate for redressing the situation (Adas 1989; Peña 1998; Bridge and McManus 2000; Fisher 2002).

This territorial takeover, discursively justified in water management and policy terms, had several important social consequences. Primarily, the public seizing of these areas translated into an important disruption of local productive systems. The expropriation of this territory resulted in the enclosure of an important part of seasonal grazing resources. In these areas local practices were excluded or tightly controlled by state representatives. In these valleys, as in many other mountainous areas of the world, local communities had developed a system of vertical and seasonal use of the different ecological niches of the mountain (Murra 1975; Netting 1981). Expropriation policies affected mid- and upper-range areas. These areas were fundamental to the yearly circulation of the herds. The top of the ranges constituted the bulk of the summer pastures. The middle mountain pastures were transition areas. Fall and spring pastures sustained the herds in their yearly transitions from the bottom of the valley in winter to the top of the ranges in summer. The loss of summer pastures increased the pressure on pastures that were supposed to sustain the herds only in winter. Herd owners had to invest in feed to compensate for the overgrazed valley fields; costs quickly escalated.

Furthermore, these policies physically removed a significant part of the network of isolated farms. Consequently, those farms that remained also lost trade partners, potential allies in communal work, and marriage possibilities. These expropriations helped to dismantle the possibilities for the social reproduction of an entire way of life, particularly since the farms were already under severe stress due to labor competition from the factories that offered what was perceived widely as an easier way of life.

At the beginning of this chapter we discussed the general economic and demographic transformations of the Valley in the last 150 years. We saw that as a consequence of the transition from the premodern agroranching valley to the modernized and industrialized mountain communities of the mid-twentieth century, most of the network of isolated farms and small villages from the ranges disappeared.

It is unquestionable that the industrial nodes located all along the river banks became highly attractive to the people living on the slopes of the Cadi Range. A closer look at the territorial consequences of the combination of water and forestry policies, however, depicts another story. Although these policies are not the sole factor in the depopulation of the farms, it is indisputable that they have played a significant role in the dismantling of those networks and the creation of a migratory movement that ultimately provided factories and mines with a much-needed labor force.

Discussion

The depicted process describes a shift in the ways nature is perceived, constructed, and managed. This chapter focused on the impact that water policies had on the way water was used and managed. Thus, this is a study mainly on governance. Water went from being a local issue to a national priority, from being privately managed to being overwhelmingly controlled by the state, and from there began a process of reprivatization. While analyzing how resources are governed, we observe that a resource has to be culturally reconstructed and extracted from the private domain and recast into the public sphere in order to be brought under the jurisdiction of national management.

While analyzing Spanish water policies, we saw that, prior to 1860, the state had paid scant attention to water management. Progressively, water entered the core of the Spanish political arena. After thirty years of timid, failed attempts, in 1900, the state and its intellectual elites took control of water by declaring it an essential national priority. Water became a central theme of the Spanish political imagination, a symbol of all the problems of Spanish society; fixing irrigation, along with deforestation, would symbolically fix a decaying society.

The "movimiento regeneracionista," interestingly enough, searched for a modernization of Spanish society. This development was not restricted to economy; it included political life and democracy. The "regeneracionistas" were enlightened intellectuals. They were liberals not inclined to trust tradition and illiterate masses. The use of modern science was the only way to rationally organize society. The state's monopoly of knowledge was also the key to power and legitimacy.

From the 1870s until the 1920s, a myriad of laws were enacted, and projects were initiated. Although their immediate, tangible effects over the Spanish landscape were not dramatic, an important hydraulic legal corpus was gradually formed. Eventually, in 1926, the Hydraulic Confederations were created. Even though the confederations initially held little or no power, their subsequent role in the management of rivers makes them very significant. They physically represented the regionalization, if not the nationalization, of water management. Each confederation had responsibilities over a whole, or a network of, watersheds.

After the 1930s, the development of more comprehensive water policies from a national perspective shifted from a theoretical priority to a tangible reality. Although the goals were obviously not the same, the *Republica* and the Franco regime invested a lot of resources and personnel in all sorts of water policies. Most dams were built between the 1930s and the early 1980s (Swyngedouw 1999, 450).

The late eighties and beyond represented a new, if not radical, change—at least of a qualitative nature in the goals pursued by water policies. This shift was marked, once again, by previous changes in Spanish culture. New uses and practices had entered with force into Spanish quotidian life. During the sixties and seventies Spain experienced a stratospheric leap in development. The end of the seventies and the eighties witnessed the emergence of environmental values and the expansion of leisure culture. Protected areas, ski resorts, and golf courses appeared all over the Spanish landscape. The state policies had to spin a little in order to incorporate these new values associated with a postindustrial society. The twentieth-century model of water use, based on modernism and development at all costs, attempted to accommodate a new set of values emphasizing a mixture of conservation, services, and leisure development.

Postmaterialistic values have emerged in urban societies worldwide (Inglehart 1997). They are connected with the contemporary social processes that urbanized rural landscapes. The prototypical postmaterialistic subject is an urban dweller searching for peace, leisure, adventure, or natural beauty. Locality and subjective positionality matter. The analysis of postindustrial landscapes, of landscapes affected by postmaterialistic values, requires careful consideration of the consequences of social and geographical scale. In the local context, the shift from extraction and transformative industries to tourism and services has to be understood as a maximization strategy designed to take advantage of a specific market situation. In other words, it does not imply an assumption of these postmaterialist values, but a rational economic alternative given the contemporary options. From a macroeconomic perspective, however, taking into account the embeddedness of the local networks with national and international socioeconomic webs and trends, the Valley has indeed shifted to a socioeconomic structure firmly connected to tourism and leisure.

The historical development of Spanish water policies involves a fundamental duality: local range versus national scope. On the one hand, we are talking about the localization of decision-making power. On the other hand, it is about discussing the origin of the needs addressed by managerial decisions at the level of the state. The eruption of water as a national issue in the late nineteenth century stripped away local jurisdiction of this resource by metaphorically connecting and turning streams and rivers in the country into national reservoirs. Consequently, decision-making power also moved from one social level to another: from local individuals and collectives in the agricultural and early industrialization era to state-regulated institutions during the consolidation phase of the modernization period. It is interesting, also, to see how the state's apparatus established significant ties to large corporations with interests in river management. Private sectors, thus, while working on hydropower generation or the provision of potable water among others, have developed some sort of shared jurisdiction with the state. However, none of these developments have evidently increased local influence on river management.

As a political issue, water changed in its meaning as human consumption priorities and environmental concerns forcefully entered into mainstream politics. Social priorities began to change as the rural areas went from a reservoir for labor, inexpensive energy, and raw materials to an environmental Eden for the enjoyment of tourists. Waterscapes turned into leisure spaces. The streams became fishing heavens where locals and urban visitors alike enjoyed trout fishing. The water reservoirs grew to be leisure spaces for kayaking, waterskiing, and swimming. At the top of the mountains, water management was also reinvented. The slopes were covered with ski slopes. Snow became a valuable commodity. Water availability was conditioned by the needs of the ski resorts.

Interestingly enough, all the land expropriated during the "protecting the dams" era became the nucleus for numerous protected areas that the state created in this part of the country. The existence of public land in the area facilitated the implementation of conservative policies and, consequently, the transformation of this landscape from industrial to natural.

Spanish landscape underwent a process of depopulation and productive stagnation in which postmaterialistic values took over as the rationale for managerial decisions (Inglehart 1997). Environmental restoration and aestheticism for tourists have become powers fueling the current economic, ecological, and social transformations.

The consolidation of democracy after 1975 perhaps did have an effect on the local possibilities for influencing political decision making at the regional level. The democratic regime, based in an open party system with responsive and democratized administrative structures, opened discussion channels for the representatives of local political institutions and organizations. Lobbying regional representatives, whether through internal party conduits or via pressure from civil society, has proven to be an efficient way to deal with local endemic problems.

The state and the corporations managing the hydropower generation industry have received enormous benefits from the appropriation and harnessing of the water resources from the mountains. This process altered in significant ways the ecology, morphology, and life conditions of the human communities in the mountains. Most of the energy and benefits produced were channeled away from the mountains to the urban centers. The decision-making process and the redistribution of production and surplus were not characterized by considerations of equity or participation. On the contrary, often the managerial rationale implemented by modern managers assumed the inadequacy of local practices and the inability of local communities to manage their resources effectively.

Modernization ideology is deeply marked by the consolidation of nation-states based on and managed according to efficiency rationales. In this kind of setting, the good of the majority clearly outweighs the needs of the few. As a result, small peripheral communities do not fair well when confronted with the demands of the urban state. This is then a debate between two concepts of equity. One concept indicates that the success of a policy may be calculated from the absolute number of citizens that benefit from it. This is the kind of rationale that, for instance, may justify rural expropriations to cover national needs. Needless to say, a policy that benefits the inhabitants of the cities will benefit larger numbers of citizens. The other concept, on the other hand, may consider that the survival of or respect for the rights of, mountain communities may deserve special treatment, via subsidies, for example. After all, many mountain communities have carried the burden of a disproportionate level of sacrifice for the sake of national energy production.

The policies studied here have been more concerned about efficiency, productivity and individualistic equity than about communitarian equity. In recent years we have seen some indices that may suggest a change in this trend. The fact is, however, that significant historical episodes of hydraulic policies in Spain have been dangerously close to pillaging processes.

This chapter also has assessed the multiple consequences that water policies have over territory and society. By looking at the local consequences on a remote area, we were able to explain the potential for territorial reverberation of these policies. Our microstudy of the Lillet Valley allows us to assess the potential collateral impacts of water policies. These collateral effects are perceptible in areas not directly geographically connected to these policies.

Indeed, water management cannot be studied as a localized, clearly delimited issue. The consequences of every hydraulic policy echo over large distances. The social and ecological footprints of dams and channels encompass entire watersheds and regions. Water policies, in the long run, were not and are not only about managing water and its associated land. They act as territorialization policies that validate natural resources appropriation and the rearrangement of both the national administrative grid and the property regimes of entire regions.

Furthermore, water policies cannot be understood without connecting productive practices, culture, and politics. Water management and political priorities are inextricably linked to the socioeconomic and cultural changes endured by local and national societies. In other words, the considerable shifts on the perception of water and of water management cannot be understood without analyzing the process of urbanization of the Spanish countryside, without paying attention to the construction of the modern Spanish state and its bureaucratic networks, and without studying the transformation of an agricultural society into a modern wage-dominated industrial society existing in a globalized world.

Water policies thus cannot be studied in isolation. As the Valley of Lillet case demonstrates, these policies hardly appear without involving a host of factors. They need to be understood as part of the repertoire of territorial policies possessed by the state. Hence, water issues need to be studied in association with other elements of the state's political activity, such as forestry, geology or energy policies. None of these state's natural resources management fields can be understood in isolation. They constitute an assemblage of sometimes coordinated, sometimes contradictory, state initiatives. All of them, in different degrees, restructure territory, natural resources, and the life of individuals.

Only a holistic approach is capable of unveiling the effects of this policy assemblage over land and people. Initially, the large territorial transformations observed in the Valley of Lillet seemed to be exclusively based on forestry policies. A detailed analysis has allowed us to connect these transformations with other water policies implemented in remote areas of the watershed, and to understand them as part of Spanish national political and ideological history. Finally, this chapter has discussed the effects of these policies on the local social systems and shown how the transformations affected not only property regimes, but productive practices and demographic patterns. In conclusion, in the Pyrenean valleys, the development of regional and state water policies did not include equity considerations in decision making.

Notes

1. Elsewhere I have dealt in detail with such a process from a territorial and political perspective (Vaccaro 2005).

2. *Cervus elaphus, Rupicapra rupicapra pyrenaica,* and *Capreolus capreolus* respectively.

3. Translated by the author.

4. During that period the legal production related to water consists in: 1860 Royal Decree, 1870 Law of Channels and Dams, 1879 Law of Waters, 1883 Law of Large Irrigation Projects and the 1905 Law of Small Irrigation Projects.

5. This period was when Spain lost its last colonies (Cuba and the Philippines). There was social distress in rural and urban areas, and a loss of economic competitiveness.

6. Decree law of March 5, 1926.

7. The prewar legislation included, for instance, the Irrigation Works Law 1932 and the Hydraulic Works National Plan 1933.

8. In the United States, President Jimmy Carter stopped federal funding for large dams in the early seventies.

9. The new Constitution 1978 proceeded to transform the centralized and authoritarian pre-1975 Spain into a decentralized state: "la España de las Autonomias."

10. Not surprisingly, the regions aligned themselves along partisan lines, taking into account the directionality of water transfers benefits.

11. In personal interviews, not corroborated by other sources, residents describe river otters living in the river near their village at the beginning of the twentieth century. River otters are considered a key biological indicator species. This is one of the first species that disappears when riverine environments degrade.

12. The village of Sant Juliá de Cerdanyola has one of such irrigation systems. Molina (2000) mentions another of these schemes in the southern face of the Cadí range between Gisclareny and Josa.

13. Colonies: in this context, industrial settlements typical of early industrialization. It was less expensive for the factories administration to rule and control a village of workers isolated and constructed ad hoc than to depend on the labor force from neighboring regular villages.

14. Other areas have a long research tradition on the history and effects of hydroelectric infrastructures; see, for instance, Blumm 1999.

15. Instituto Nacional para la Conservación de la Naturaleza (National Institute for the Conservation of Nature).

References

Abramson, A., and D. Theodossopoulos, eds. 2000. *Land, Law, and Environment: Mythical Land and Legal Boundaries*. London: Pluto Press.

Adas, M. 1989. *Machines as the Measure of Men: Science, Technology, and the Ideologies of Western Dominance*. Ithaca: Cornell University Press.

Agrawal, A., and K. Sivaramakrishnan, eds. 2003. *Regional Modernities: Cultural Politics of Development in India*. Delhi: Oxford University Press.

Arqué, M., A. Garcia, and X. Mateu. 1982. "La Penetració Del Capitalisme A Les Comarques del'Alt Pirineu." *Documents d'Analisi Geografica*, no. 1, pp. 9–67.

Arrojo, P., et al. 2004. *El Agua en España: Propuestas de Futuro*, Madrid: Ediciones del Oriente y del Mediterráneo.

Bauer, E. 1980. *Los Montes de España en la Historia*. Madrid: MAPA.

Bauer, K. 2004. *High Frontiers: Dolpo and the Changing World of Himalayan Pastoralists*. New York: Columbia University Press.

Blatter, J., and H. Ingram. 2001. *Reflections on Water: New Approaches to Transboundary Conflicts and Cooperation*. Cambridge, MA: MIT Press.

Blumm, M. 1999. "The Northwest's Hydroelectric Heritage." In D. Goble and P. Hirt, eds., *Northwest Lands, Northwest Peoples*. Seattle: University of Washington Press, pp. 264–294.

Boneta, M. 2003. *La Vall Fosca: els llacs de la Llum. Desenvolupament Socioeconomic a Comencament del Segle XX*. Tremp: Garsineu Edicions.

Braun, B. 2002. *The Intemperate Forest: Nature, Culture and Power in Canada's West Coast*. Minneapolis: Minnesota University Press.

Brenner, N. 1997. "State Territorial Restructuring and the Production of Spatial Scale: Urban and Regional Planning in the Federal Republic of Germany, 1960–1990," *Political Geography* 16, no. 4: 273–306.

Bridge, G., and P. McNaus. 2000. "Sticks and Stones: Environmental Narratives and Discursive Regulation in the Forestry and Mining Sectors," *Antipode* 32 no. 1:10–47.

Brown, L., and Helen Ingram, eds. 1987. *Water and Poverty in the Southwest.* Tucson: University of Arizona Press.

Casals, V. 1996. *Los Ingenieros de Montes en la España Contemporánea: 1848–1936.* Barcelona: Serbal.

Castells, M. 1996. *The Rise of the Network Society.* Malden, MA: Blackwell Publishers.

Castells, M. 1997. *The Power of Identity.* London: Blackwell Publishers.

Costa, J. 1975 (1892). *Política Hidráulica y Misión Social de los Riegos en España.* Madrid: Edición de la Gaya Ciencia.

Cronon, W. 1991. *Nature's Metropolis: Chicago and the Great West.* New York: Norton.

Donahue, J., and B.R. Johnston, eds. 1998. *Water, Culture, and Power: Local Struggles in a Global Context.* Washington, DC: Island Press.

Fisher, F. 2002. *Citizens, Experts, and the Environment: the Politics of Local Knowledge,* Durham: Duke University Press.

Gandy, M. 2002. *Concrete and Clay: Reworking Nature in New York City.* Cambridge, MA: MIT Press.

Goble, D., and P. Hirt. 1999. *Northwest Lands, Northwest Peoples,* Seattle: University of Washington Press.

Gómez, J. 1992. *Ciencia y Política de los Montes Españoles (1848–1936).* Madrid: ICONA.

González, R. 2001. *Zapotec Science: Farming and Food in the Northern Sierra of Oaxaca.* Austin: University of Texas Press.

Hannah, M. 2000. *Governmentality and the Mastery of Terrirory in Nineteenth-Century America.* Cambridge, UK: Cambridge University Press.

Hicks, G., and D. Peña. 2003. "Community *Acequias* in Colorado's Rio Culebra Watershed: A Customary Commons in the Domain of Prior Appropriation." *University of Colorado Law Review* 74, no. 2:387–486.

Hundley, N. 1975. *Water and the West: the Colorado River Compact and the Politics of Water in the American West.* Berkeley: University of California Press.

Hundley, N. 1992. *The Great Thirst: Californians and Water, 1770s–1990s.* Berkeley: University of California Press.

Inglehart, R. 1997. *Modernization and Postmodernization: Cultural, Economic and Political Change in 43 Societies.* Princeton, NJ: Princeton University Press.

La Calle, A. 2004. "El Nuevo Marco Jurídico que Introduce la Directiva Marco de Aguas en la UE." In P. Arrojo et al. *El Agua en España: Propuestas de Futuro.* Madrid: Ediciones del Oriente y el Mediterráneo, pp. 69–123.

Maluquer de Motes, J. 1998. *Historia económica de Catalunya: segles XIX i XX.* Edicions de la Universitat de Barcelona i Proa: Barcelona.

Mangas, J. 1984. *La Propiedad de la Tierra en España.* Madrid: Instituto de Estudios Agrarios, Pesqueros y Alimentarios.

Molina, D. 2000. *Conservació I Degradació De Sòls a les Àrees de Muntanya en Procés D'abandonament. La Fertilitat del Sòl al Parc Natural del Cadí-Moixeró.* Master's Thesis in Geography. Universitat Autonoma de Bellaterra.

Murra, J. 1975. *Formaciones Económicas y Políticas del Mundo Andino.* Lima, Pontificia Universidad Católica del Peru.

Nel.lo, O. 2001. *Ciutat de ciutats.* Barcelona: Empuries.

Netting, R. M. 1981. *Balancing on an Alp: Ecological Change and Continuity in a Swiss Mountain Community.* New York: Cambridge University Press.

Orlove, B. 2002. *Lines in the Water: Nature and Culture at Lake Titicaca.* Berkeley: University of California Press.

Ortega, N.1979. *Política Agraria y Dominación del Espacio,* Madrid: Editorial Ayuso.

Peña, D. 1998. *Chicano Culture, Ecology, Politics: Subversive Kin.* Tucson: University of Arizona Press.

Pérez, M. T. 2004. "De Costa al Paradigma del Desarrollo Sostenible. Claves Históricas de Una Crisis." In Arrojo, P., et al. *El Agua en España: Propuestas de Futuro.* Madrid: Ediciones del Oriente y el Mediterráneo, pp. 125–152.

Rivera, J. 1999. *Acequia Culture: Water, Land, and Community in the Southwest.* Albuquerque: University of New Mexico Press.

Saberwal, V. 1999. *Pastoral Politics: Shepherds, Bureaucrats, and Conservation in the Himalaya.* Delhi: Oxford University Press.

Sauri, D., and L. del Moral. 1999. "Changing Course: Water Policy in Spain," *Environment* 41, no. 6:31–36.

Sauri, D., and L. del Moral. 2001. "Recent Developments in Spanish Water Policy. Alternatives and Conflicts at the End of the Hydraulic Age," *Geoforum* 32:351–362.

Shore, C., and S. Wright, eds. 1997. *Anthropology of Policy: Critical Perspectives on Governance and Power.* London: Routledge.

Strang, V. 1997. *Uncommon Round: Cultural Landscapes and Environmental Values.* Oxford: Berg.

Strang, V. 2004. *The Meaning of Water.* Oxford: Berg.

Swyngedouw, E. 1999. "Modernity and Hibridity: Nature, *Regeneracionismo,* and the Production of the Spanish Waterscape, 1890–1930." *Annals of the Association of American Geographers* 89, no. 3:443–465.

Swyngedouw, E. 2001. "Scaled Geographies: Nature, Place, and the Politics of Scale." In R. McMaster and E. Sheppard, *Scale and Geographical Inquiry: Nature, Society and Method.* Oxford: Blackwell.

Tarraubella, X. 1990. *La Canadenca al Pallars.* Tremp: Garsineu Edicions.

Vaccaro, I. 2005. "Property Mosaic and State-Making: Governmentality, Expropriation and Conservation in the Pyrenees." *Journal of Ecological Anthropology* 9:4–19.

Vandergeest, P., and N. Peluso. 1995. "Territorialization and State Power in Thailand," *Theory and Society* 24:385–426.

Wickham, G., and G. Pavlich. 2001. *Rethinking Law, Society, and Governance: Foucault's Bequest.* Oxford: Hart Publishing.

Williams, R. 1972. *The City and the Country.* Oxford: Oxford University Press.

Worster, D. 1985. *Rivers of Empire: Water, Aridity, and the Growth of the American West.* New York: Pantheon Books.

9

Whose Water Is It Anyway? Water Management, Knowledge, and Equity in Northeast Brazil

Maria Carmen Lemos

Water management has historically posed exceptional challenges of equity and justice to policymakers around the world. Lately, the potential exacerbation of water scarcity and water-related disasters because of climate variability and the potential impacts of global climate change has upped the stakes for the vulnerable poor who are expected to endure the brunt of its negative consequences (Kates 2000; Adger et al. 2003). Because they are less able to mitigate and adapt to the negative consequences of water regime change, less developed countries have become the focus of attention and the locus of policy experiments that seek simultaneously to improve short-term water management efficiency and to promote long-term sustainability and adaptability.

Increasingly, albeit slowly, water management has moved from a mostly technical and elite-dominated affair to a process where decentralization, privatization, and stakeholder participation have become prevalent goals (Gleick 2000; Ribot 2002). This new paradigm of water management has sought to design and implement institutions that are democratic, economically viable, responsive to future change, and lead to long-term sustainability. Efforts to implement water management reform are going on in places as diverse as Chile, South Africa, Mexico, and Brazil with different levels of breadth, commitment, and achievement (Wester, Merrey, and deLange 2003; Bauer 2004; Brannstrom, Clarke, and Newport 2004; Lemos and Oliveira 2004). Water management reforms—in accord with the Dublin Statement of 1992[1]—have been heavily promoted by organizations such as the World Bank, the Organisation for Economic Cooperation and Development, and the United Nations.

In this chapter, I argue that while the emergence of a new paradigm of water management in Brazil, including the creation of stakeholder river basin councils, has opened the door for broader societal participation, it is the availability and accessibility of knowledge that has the highest potential to make this participation effective. In other words, although formal institutional rules are a necessary condition for participation, they are by no means sufficient to foster effective participation, both in terms of outcome (i.e., more efficient and equitable management) and process (i.e., more democratic, transparent, and accountable decision making). Moreover, in the context of Brazil's dysfunctional, insulated policymaking—in which water agencies' decisions may be grounded simultaneously in technical criteria and vulnerable to political interference—the control of knowledge production and knowledge use has had deep implications for the equitable distribution of resources.

Around the world, water management reforms have ranged from the privatization of water supply and basic sanitation to the implementation of full-fledged water markets. Two aspects of this effort speak directly to equity issues. The first is a push to reframe water as a common good with economic value, that is, a good for which users should pay. While water tariffs to recover operational and management costs of water supply and basic sanitation have been conventionally applied to consumers for generations, the new paradigm focuses on bulk water charging, sparking a debate among users, policymakers and a broad range of stakeholders—especially environmental non governmental organizations (NGOs). In principle, if users are made to pay for a good that they have customarily been able to access free of charge, they will be more likely to conserve it. However, the manner in which pricing schemes are negotiated, set, and implemented has a profound impact on the equity of distribution, access, and costs for different classes of users.[2]

The second aspect of water management with deep equity implications, and the focus of this chapter, is the push for the creation of participatory institutions, such as stakeholder councils in decision arenas that were previously dominated by the possessors of technical expertise. A common assumption behind the expectation of successful stakeholders' councils is that participation *matters* both in terms of outcomes and processes. That is, stakeholders' involvement contributes to solving the problems that brought them together and promotes desired practices such as democratization, transparency, and accountability in the context of policymaking. Yet, empirical evidence on the ability of stakeholder partnerships to reach these goals has been mixed (Leach, Pelkey, and Sabatier 2002; Manor 2005).

Scholars have identified a variety of reasons behind such failures ranging from the insincerity of some participatory schemes—for example, participation that is meant to "legitimize" certain policies in the eyes of funding agencies—to the unequal organization of stakeholder groups and the difficulty of including a representative array of stakeholders in the councils (Mohan and Stokke 2000; McEwan 2003; Wester, Merrey, and deLange 2003; Abers and Keck 2005). These problems have appeared in the case of water councils, where members have not participated effectively, key social groups—especially the poor—are not effectively represented, and councils are unable to define and carry out agendas for action (see, for example, Wester, Merrey and deLange 2003; Lemos and Oliveira 2004; Formiga-Johnsson and Kemper 2005). Equity implications of stakeholder involvement include tradeoffs between different users (large vs. small, upstream vs. downstream, public vs. private, current vs. future users), uses (e.g., irrigation, shrimp farming, recreation, water supply, etc), and type of problem (pollution vs. scarcity).

At its best, stakeholder involvement may increase the level of equitable allocation, democratization, accountability, and transparency of water management. At its worst, it may threaten resource sustainability and equitable distribution by allowing a few elite actors to "capture" the decision-making process at the expense of other stakeholders. In this process, the use of technoscientific information has the potential to shape both policy outcome and policy process.

On the one hand, knowledge can contribute to more effective management by informing stakeholders about system capacity and fluctuations, potential disruptions to resource availability (e.g., drought or flooding), implications of intra- and inter-basin water transfers, long-term availability, and intergenerational implications of different levels of resource use (i.e., climate change impact scenarios). It can also inform stakeholders about the implications of water quality for current use and future sustainability of water resources, and support decisions regarding water zoning plans and pricing schemes. Moreover, the ability to transfer knowledge and adopt innovation is an essential factor in building adaptive capacity to climate variability and change (Smit et al. 2000). In this sense, knowledge can potentially improve effectiveness and democratize decision making since better-informed stakeholders can make better-informed decisions.

On the other hand, if controlled by a few actors seeking to bolster their position vis-à-vis other stakeholders, knowledge can insulate decisions and exacerbate power imbalances between those with access to knowledge and those without that access (Lemos 2003). In such cases, knowledge can have critical implications for the "elite capture" of the decision-making processes, which in turn can affect issues of equity and justice in water management. Here, the difference between democratization and insulation rests on the rules of engagement of stakeholders and the practices regarding the availability and accessibility of knowledge.

In the next sections, I discuss an experience of technoscientific knowledge use in water negotiation and allocation using data collected in the context of ongoing field research in the Lower Jaguaribe Banabuiú River Valley (LJBV) in Ceará, Northeast Brazil.[3] In this valley, a water user's commission (Comissão de Usuários) has been actively negotiating reservoir water allocation for the past ten years. To support water allocation meetings, the Water Resources Management Company (Companhia de Gestão dos Recursos Hídricos—COGERH)—Ceará's state water management agency—routinely puts together reservoir scenarios to inform different groups of stakeholders (including water users) of the projected availability of water for upcoming seasons. Once a year, stakeholders and COGERH técnicos meet and, based on the scenarios, they: (a) debate different potential amounts of water volume, water loss (e.g., through evaporation), and discharge, and their consequences for short and mid-term water availability (which in turn depends on the likelihood of drought); (b) negotiate between different kinds of uses and needs for the allocation of water resources; and, (c) try to build consensus on the volume of water that will be available to be discharged from the valley's two main reservoirs.

The implications of the negotiated allocation process to the issues of the equity of water distribution in the Lower Jaguaribe Banabuiú River Valley are twofold. First, the extent to which different kinds of uses and interests (including intergenerational) are represented or not in the context of the users' commission has a profound effect on the level of equitable distribution of resources and on the sustainability of the system. While the current allocation model seems to be significantly more participatory than previous management schemes in Ceará (Lemos and Oliveira 2005), the level of inclusion and representation of stakeholders has recently been the subject of closer scrutiny (Ballestero 2004; Taddei 2005).

What new evidence shows is that despite more participation, the inclusion of nonelites—such as small farmers, rural workers, and rainfed farmers—has been thwarted both in terms of representation (they are less represented) and influence (they exert less influence during the allocation meetings). Second, the extent to which the use of technical knowledge either democratizes or further insulates decision making shapes the level of access of different stakeholders not only to effective decision making but also to water. Here, despite the effort from local técnicos to improve communication and availability, there is evidence that a substantial number of stakeholders find technical information neither available nor accessible. Moreover, there is a widespread perception of técnicos as the most powerful actors in the water management process, more powerful than either economic or political elites.

Next, I will briefly examine these issues focusing on the equity implications of knowledge use in the context of Brazil's water reform. Sections 2 and 3 respectively will describe the patterns of technical information use in policymaking and the water management institutional environment in Brazil. Section 4 examines decision-making strategies, the use of technical knowledge in water management in Ceará, and their implications for equity issues. I conclude with a few preliminary findings and suggest new areas for further research.

Technocrats and Politicians: Policy, Insulation, and Accountability

Historically, technical decision making has dictated water management in Brazil. Especially in Northeast Brazil, but also in other regions, those trying to solve water-related problems provided solutions that continuously upgraded increasingly complex infrastructure systems. This approach to water-related problems was firmly grounded in a technocratic and exclusionary decision model that often shaped policymaking in Brazil (Schmitter 1971). The technocratic tradition goes back to the 1930s when, as a consequence of the modernization of the state, a strong bureaucracy emerged based on the multiplication and expansion of both public and private organizations. As I have argued elsewhere,

While in the classic Weberian model politicians and bureaucrats play very distinct roles, that is, politicians make policy decisions and bureaucrats implement them, in Brazil, the line between politics and bureaucracy has been purposely blurred under the guise of improving "efficiency" in policy-making. The underlying assumption was that politicians, because of their vulnerability to electoral politics, might fall prey to special interests and clientelistic relationships, which, in turn, could lead to biased policy decisions. Bureaucrats, on the other hand, because they are bound by their expertise and, in principle, should have no political agenda, are much more qualified to make the "best" policy decisions and implement them efficiently. (Lemos 2003, 109)

Hence, throughout Brazil's republican history, but especially in the 1960s and 1970s, political leadership (both democratically elected and authoritarian) attempted to insulate bureaucratic systems as a strategy to foment development. By singling out some agencies and providing them with financial and human resources unavailable to the bulk of the "bureaucracy," these leaders expected insulated technocracies to perform at a higher level of competency than other sectors of the government. Insulated agencies attracted high-quality professionals by offering market competitive wages and fringe benefits, by adopting strict meritocratic selection and promotion processes, and by "protecting" their decision making from traditional political meddling. The technocracy differed from traditional bureaucracy to the extent that its performance depended on specific technical and professional expertise. Most importantly, they operated from decentralized agencies (public and mixed enterprises and autonomous entities) that were relatively protected from practices such as clientelism, nepotism, the spoils systems, and corruption (Nunes and Geddes 1987).

On the one hand, insulation contributes to effective implementation of policy because it preserves material and human resources and the commitment necessary to implement reform (Geddes 1990). This, in turn, increases autonomy, and hence, state capacity. Still, "capacity-enhancing reforms . . . occur only when the political leaders who must initiate them can expect to benefit from the reforms enough to outweigh the cost of losing the electoral advantages provided by the distribution of patronage" (Geddes 1990, 218). Therefore, it is not surprising that the most encompassing period of bureaucratic insulation in Brazil coincides with that of authoritarian rule when the military and its appointed governors were not vulnerable to election results.

On the other hand, insulated technocracies operate virtually unchallenged. Technocratic decision making may defy basic precepts of democracy by limiting the number of participants and policy alternatives and rendering technocrats unaccountable to elected officials and clients (Etzioni-Havely 1983). Not surprisingly, tensions between insulation and accountability have had a lasting, mostly negative, effect on Brazil's democracy and the equitable distribution of resources (Reis 1990). Even after the transition to democracy, many technocratic agencies were able to maintain their legitimacy by articulating their authority in terms of their technical expertise (Lemos 2003).

However, the level of insulation achieved by technocratic agencies has varied significantly across geography and time. While a few federal agencies were able to protect their integrity for the most part (Nunes and Geddes 1987), other agencies, especially at lower scales of government, were subject to a combination of insulation and political meddling that mirrored the broader policymaking environment in Brazil. In other words, even within insulated agencies, the level of political meddling would vary with specific policy areas and through time. For example, within the same agency, some projects would be more insulated than others. The election of a new government could bring serious threat to levels of insulation even if the agency's goals and technical cadre remained relatively untouched (Lemos 1998; Lemos and Oliveira 2004).

This pattern of dysfunctional insulated policymaking has historically shaped water management in Brazil. Decisions were both firmly grounded in the technocratic model and at the same time vulnerable to the interference of politics and outside actors whose agendas did not necessarily correspond with what técnicos had determined to be the "best possible solution" (Lemos 2003). In Northeast Brazil, for example, the implementation of the infamous "hydraulic solution"—that is, policies that favored the construction of reservoirs and canals to store and transport water to deprived areas—was attractive because it simultaneously met technocrats' ideas of technical progress, provided politicians with the opportunity to accrue political capital (through both pork-and-barrel and clientelitic distribution of benefits), and met the interests of large contractors who stood to be retained as service providers. For example, in Ceará alone, some 7,000 reservoirs were built with public resources. While a few were massive public works projects planned to normalize water flows and redistribute water across the region, the majority were located on private property where public access and the benefits were limited for those who were the most vulnerable to water shortage (Garjulli 2001).

While, ideally, technical knowledge can allow for more transparent and better-informed decision making, in Ceará (and in other regions of Brazil), it has insulated decisions and alienated stakeholders. Access to and use of knowledge in Ceará was not equal. It favored those with power relationships and institutions. As technical analysis becomes more prominent than other informational input (including opinions and interests of nontechnical sources), it may "squeeze out other forms of information, decision-making routines, and claims" (Healy and Ascher 1995, 13). Indeed, when trying to gain political advantage, groups may be tempted to exaggerate or distort information when that information serves to support the interests of one group over another.

Water Reform and Institutional Change

In the 1990s, the Brazilian government initiated a reform process that replaced the previous centralized, top-down and sectoral system with a decentralized, participatory, and integrated model that adopts the river basin as the main unit of water management. While a few states instituted the reform as early as 1992, it was not until 1997 that the federal government enacted Law 9,433 also known as the "Water Law" (Lei da Águas). It instituted the National Policy for Water Resources and created the National System for the Management of Water Resources. It also created a National Water Agency (Agência Nacional de Água—ANA) that both oversees the application of the law and has jurisdiction over the management of interstate river basins.

Despite quite different rates of implementation across Brazil's twenty-six states and numerous river basins, the new legislation has "shaken up" water management nationally. The new legislation introduced management mechanisms more in tune with the democratization of state-society relations following the demise of the Brazilian military dictatorship in the mid-1980s. It was also more in accord with the new water management paradigm spelled out in the Dublin Principles. These included (with significant variations across states and river basins both in terms of institutional design and degree of success): (a) the organization of management at the river basin level, overhauling a previous system that favored state and federal jurisdictions; (b) the creation of specific regulation to protect water resources at the river basin level; (c) the decentralization of decision making and resources, which included the design of a new system of water-use permits and charges and the creation of different scales of public participation such as the organization of river basin-level councils and State and National Water Councils; (d) the insertion of water resources management within a larger realm of environmental concerns that challenged the traditional supremacy of economic criteria in the management of water; and (e) the understanding of water as a public good—but also an economic good—for many, the most debatable and controversial aspect of the new legislation.

One particularly novel aspect of the law has been the creation of basin-level councils whose tripart membership represents users, organized sectors of civil society, and the state, although the proportion of each sector varies widely across states and river basins. River basin councils' responsibilities vary considerably across basins and include designing and implementing bulk-water permit and charging systems, approving river basin management and water zoning plans, and facilitating conflict resolution among users.

The Brazilian system, although based on national regulation, has followed a highly federalized model that has afforded much flexibility to states to design and implement institutions that better fit the characteristics of their water resources and sociopolitical systems. Thus, different states have pursued different strategies and created different structures to manage their water resources. The state of Ceará, for example, created a separate Users Commission that work in parallel with the more "official" River Basin Committee. This has not only increased the opportunity for users' participation, but also has been hailed as a model of successful river basin–level management (ANA 2002). CEIVAP, the federal committee created in the Paraíba do Sul River, in turn, has managed to implement a user-fee system that is funding the basin's operational agency (Formiga-Johnsson, Kumler, and Lemos 2007).

With 90 percent of its territory located within the Brazilian semiarid, and with an average rainfall of 400 mm in the hinterland and 1,200 mm in the coast, Ceará is among the driest and poorest states in Northeast Brazil (Lemos et al. 2002). Traditionally, water policy and drought planning have been highly politicized and closely tied to the region's infamous "drought industry," that is, the appropriation of drought-relief funds for private gain. As mentioned earlier, in Ceará, part of the problem was an antiquated water management system that traditionally had focused on infrastructure building and had often been used as political currency. The situation changed in the early 1990s when ahead of most Brazilian states and even before the federal government, the state of Ceará started to reform its water management. As part of a series of reform-oriented state administrations (Tendler 1997), and in response to a long period of drought, which threatened water supply to the capital city of Fortaleza, a concerted effort was directed to design a new set of institutions to manage the state's water resources (Lemos and Oliveira 2004). This included the hiring of expert consultants as well as the study of state-of-the-art management options being implemented in other parts of the world.

At about the same time, the government of Ceará approached the World Bank with a proposal for the Bank to finance new water infrastructure, including the construction of new reservoirs in areas not covered by the existing network (Kemper and Olson 2001). The Bank agreed but insisted on a few conditions. The first condition was that the state implement and use the instruments outlined in the new law, including the creation of river basin committees and the introduction of tariffs for all water users (including irrigation). The second condition was that the state create a water resources management company (Kemper and Olson 2001, 342). As a result, COGERH was created in 1993, with financing from the World Bank as an attempt to avoid the common pitfall of "paper laws," that is, reformist legislation doomed to failure because of lackluster or inadequate implementation. COGERH's responsibilities include the monitoring and management of Ceará's water resources, maintenance and operation of the state's water infrastructure, and the implementation of the institutions of the new water resources law including the organization of users across the state's eleven river basins.

COGERH followed Brazil's traditional model of insulated technocracy in which political leadership purposefully insulated agencies—created under the guise of technical expertise, meritocratic hiring, and plenty of resources—from the maladies of the inefficient public sector associated with third-world bureaucracies. However, in one aspect, COGERH was critically different from other technical agencies: at the insistence of outside consultants, the agency included social scientists in addition to the usual makeup of engineers and hydrologists associated with water management agencies. Thus, "the combination of social and physical scientists within the agency allowed for the amalgamation of ideas and technologies that critically affected the way the network of technocrats and their supporters went about implementing water reform in the state" (Lemos and Oliveira 2004, 2127). In the mid-1990s, COGERH started to put together reservoir scenarios to support water management. These scenarios became a valuable tool in supporting stakeholders to negotiate the allocation of water resources among different users.

In this context, new ideas and technologies may work as a critically enabling tool. Indeed, in the case of COGERH, technical knowledge was instrumental not only in informing the creation of many of the organizational schemes pursued by COGERH técnicos but also in inviting mobilization from users who perceived their participation as meaningful and effective. One particularly innovative aspect of COGERH's organization of management at the river-basin level was the creation of Users' Commissions that—in addition to the more institutionalized river basin committee envisaged in the legislation—would be able to participate directly in water allocation decision making. As mentioned earlier, the Commission meets periodically to evaluate and plan for water use at the river-basin level and functions in parallel to the river basin committee. It is also within the context of the Commission that technoscientific information has been used to inform water allocation decisions among different users, especially irrigated farms, large agribusinesses, and water utilities. After ten years and an effective decrease in the level of water use and conflict (Formiga-Johnsson and Kemper 2005), the Ceará case is hailed as the model to be followed in Northeast Brazil, and possibly other semiarid regions of the world.

Management, Equity, and Technical Knowledge

Similar to other areas of the world, water systems in Brazil pose significant challenges to policymakers in equity terms: (a) they are subject to multiple, sometimes contradictory uses; (b) a variety of users depend on water resources at diverse levels of need; (c) management systems span different scales, and many times have overlapping jurisdictions; and (d) resources themselves are subject to different levels of stress, scarcity, and depletion (Rayner et al. 2002). In the context of water scarcity prevalent in Northeast Brazil, several equity issues rest at the heart of water management:

• Conflict between different kinds of uses, including irrigation, fisheries, water supply, basic sanitation, and industry. Within these user categories, there is further conflict between large and small users. For example, in irrigation, there is a conflict between agribusiness, medium and small irrigation perimeters, and rainfed agriculture that relies on humid areas around reservoirs in which to plant. Although all these groups depend on reservoir water, their needs in terms of volume and timing of discharge vary (see next section).

• Intrabasin transfer and resources distribution. In Ceará, since water is scarce and badly distributed (the Jaguaribe is the only "normalized" river in the state), the state built a complex system of reservoirs and canals to reallocate water in the region. For example, the capital city of Fortaleza, the largest consumer in the state, depends on transfers from the Jaguaribe to supply its residents and industrial users. The metropolitan basin of Fortaleza is also the biggest generator of revenues from bulk water charges in the state, virtually financing the water management system in place—including COGERH. Thus, while users from Lower Jaguaribe Banabuiú River Valley may resent the transfer to Fortaleza, the system would collapse without the funds it generates.

• Water quality and environmental impacts. In Ceará, declining water quality, especially as a consequence of untreated sewage from urban areas and pollution from pesticides and shrimp farming, has exacerbated concern over the future sustainability of the basin. However, the environmental implications of these changes have not been included in the state's policy agenda, and their long-term effects are not known (Lemos and Oliveira 2004; Formiga-Johnsson and Kemper 2005). Another issue with potential important environmental and equity implications is groundwater use, which has been virtually unregulated in the past and has been mostly ignored in the new water management law.

• Institutional design and fit. The current institutional design, albeit more participatory, may still skew water distribution by allowing large users to request water permits outside the jurisdiction of the River Basin Committee or Users Commission. Currently, any user can obtain a permit (and pay for the bulk water) without the approval of either council. Although, for the most part, conflict between permit holders and other users has not arisen, the lack of a requirement for prior approval in the formal institutional design has been a sore point with stakeholders (Lemos and Oliveira 2004). This situation puts large users at a clear advantage over small users.

Moreover, many of these issues cross over and overlap, adding complexity. For example, shrimp farmers in Ceará are not the largest consumers of water but are resented by other stakeholders because of the negative impacts of their farms on the environment (after usage, water is returned to the river untreated). Irrigation, in turn, albeit the highest consumer (47.1 percent), is the most resistant to water charges (Formiga-Johnsson and Kemper 2005). Finally, despite the potential for inequity—since large users can apply for water permits outside the purview of the Committee and Users Commission—it is precisely these large paying users who finance the management system.

Reform, Decentralization, and Knowledge

Within the rather loose institutional framework of Brazil's water reform, policymakers are afforded greater degrees of freedom, not only to create new institutions, but also to change the existing ones. Hence, "within the constraints imposed by particular technological or economic configurations, actors can modify institutions to solve new problems, to facilitate network-based collective learning, or to achieve increasing efficiency" (Clemens and Cook 1999, 451). This seems to be exactly what COGERH técnicos tried to do by opening the decision-making process at the river basin level, albeit with different levels of commitment and breadth of perspective.

Ideas and knowledge have played a pivotal role in this process of institution building in Ceará. First, ideas—here defined as a cluster of beliefs affecting the design of strategies of action geared toward policy outcome—shaped the creation of the new water management structure by supporting the initiative of policymakers to push for decentralization and participation. As argued before, the inclusion of social scientists in COGERH and the creation of a specific department to organize users within the agency changed the dynamics of reform to an unprecedented level of participation in water allocation. Many of these técnicos have been active in reformist social movements and politics for years. Their belief systems and worldviews heavily influenced their actions (Lemos and Oliveira 2004).

In the Lower Jaguaribe-Banabuiú river basin, the implementation of participatory councils went further than the suggested framework of River Basin Committees to include the Users Commission to negotiate water allocation among different users directly. Técnicos specifically created the Commission independently of the "official" state structure to emphasize their autonomy vis-à-vis the state (Lemos and Oliveira 2005). This agenda openly challenged a pattern of exclusionary and clientelist water policymaking prevalent in Ceará. In practical terms, these changes meant a substantial departure from the way water allocation was negotiated in the past. The ability of these técnicos to implement the most innovative aspects of the Ceará reform can be explained partly by their insertion into policy networks that were instrumental in overcoming the opposition of more conservative sectors of the state apparatus and their supporters in the water user community (Lemos and Oliveira 2004).

The Users Commission meets once a year (with smaller meetings occurring in between) to negotiate bulk water allocation. A larger pool of stakeholders elects representatives from users, the state, and organized civil society to participate in the negotiated allocation process. Membership is broken down as follows: (a) twenty-seven representatives from the municipal government (25 percent); (b) eighteen representatives from the state and federal governments (17 percent); (c) thirty-two representatives from civil society (30 percent); and (d) thirty representatives from the sectors of water users (28 percent) (Taddei 2005).

Although there is some variation in the electoral process from year to year, there is evidence of active negotiation and bargaining between members to get a seat (Ballestero 2004, personal interviews). The level of representation within the Commission is often questioned on two fronts. The first front is that membership is biased toward those with greater resources (both material and social), who are able, not only to keep high rates of participation in User Commission meetings, but also are able to mobilize the political and social capital that is necessary to be elected. For example, one explanation for the lower level of participation among the poor is their lack of resources to travel to the preparatory and allocation meetings (Taddei 2005). Second, representation is thwarted by the lack of accountability between members and the constituencies they are supposed to represent, since both the level of previous consultation and reporting back to the constituency is low among members (Ballestero 2004; Taddei 2005).[3] Furthermore, although the meetings are public and therefore open to all, in case the Users Commission fails to reach consensus on water allocation, only "official" members can cast a vote.

In their annual meeting, técnicos from COGERH prepare a series of reservoir management scenarios that includes amount of available water, rates of evaporation, and other specific conditions affecting water availability. Commission members, led by these técnicos, debate alternative discharge scenarios based on water availability and users' needs. Not surprisingly, although state and society also have representatives in the Commission, users exert the biggest influence in the allocation decisions (Ballestero 2004). In this situation, there is the possibility that users' lower risk-averseness may bring the system to a higher level of stress, that is, in a "tragedy of the commons scenario," users will overconsume water at the expense of long-term sustainability. In addition, many believe that because the current system does not take into account intergenerational environmental issues and the long-term sustainability of the river basin, the state needs to step in to protect the stake of both society and ecosystems. With the exception of a few attempts to bolster environmental education, the state's River Basin Committees and User Commissions have sidestepped the issue of environmental impact and not considered long-term water sustainability in terms of future consumption (Formiga-Johnsson and Kemper 2005). To guarantee short-term sustainability and prepare for the eventuality of a multi-year drought, COGERH técnicos have attempted to include some safeguards in the negotiated allocation process. For example, users are allowed to decide reservoir discharge up to a predetermined level. During meetings, técnicos often use their influence to advocate decisions that are more conservative. Although there is a risk that direct user input in water allocation would lead to overuse, this has not so far been the case in the Jaguaribe/Banabuiú (Lemos and Oliveira 2004; Formiga-Johnsson and Kemper 2005). The reasons can be traced to existing conflicts along three dimensions: (a) the presence of multiple users with conflicting interests; (b) the fact that different amounts of water have to be released from the three major reservoirs to meet users' needs; and (c) the tradeoffs between users from the lowland and highlands of the basin.

Since human consumption is a priority, large users, such as urban water supply companies, will push for lower levels of discharge to ensure longer periods of water availability. Water systems that depend on intrabasin transfer favor more conservative rates of release as well. Irrigation farmers, in turn, have an incentive to maximize water consumption as soon as possible to guarantee the viability of their economic activity in the short run. Yet they are divided by geography, since the conflict between users from the basin's lower and higher lands also helps to keep water discharge in check. Thus, for users in the lowlands, it is better that larger amounts of water are released each season to increase their planted area (the area around the reservoirs that is naturally irrigated as the level of water recedes). For users in the highlands, it is more advantageous that the level of the reservoirs remains higher to supply their irrigated farms as needed throughout the growing season. Similarly, irrigators' downstream push for more release while irrigators upstream push for less. Finally, the fact that técnicos build scenarios for the whole system, but water is released from different reservoirs within the basin, can affect users differently depending on their location in the basin. Therefore, there is an incentive for some users to protect resources in their surrounding area as they negotiate the amounts discharged from specific reservoirs as part of the broader allocation system (Lemos and Oliveira 2004, 2129).

The main implication for equity of the negotiated allocation is the fact that a greater number of both large and small users are effectively participating in the process. However, the degree of democratization of decision making and its impact on equitable distribution of resources is far from straightforward. Whether there is a positive impact depends on one's point of view. On the one hand, participation can strengthen and legitimize the new water reform institutions, expand the array of stakeholders deliberating about water use and sustainability, and strengthen the capacity of reform-oriented policy networks to push for further reform (Lemos and Oliveira 2004). Moreover, instead of resorting to clientelism to push for the specific agendas, users now can participate in a much more transparent fashion. In consequence, the system has grown substantially more accountable to stakeholders than in the past. In practical terms, this has meant that conflicts have been better addressed between different uses and between large and small users. The main implication of the use of technoscientific knowledge for equity is that this participation is more meaningful and effective because stakeholders are able to make better-informed decisions when they have access to this knowledge. Thus users, who would not have had a say in the past, directly or indirectly, now have the opportunity to defend their interests using expert knowledge instead of only acquiescing to what "experts" have to say. New knowledge and technologies such as reservoir modeling, not only provided for better-informed decisions within the new management schemes, but also for more active participation by users.

In the context of Ceará's Users Commissions, the advantages in this case are many. First, users are more likely to abide by the decisions at the river-basin level (at this point there is not an established enforcement system, so basically, social pressure is the only weapon the técnicos and other users have to enforce how much water is being used) since they have been directly involved in the decision-making process. Second, by making simplified reservoir models available to users, COGERH is not only enhancing knowledge about the river basin but also is crystallizing the idea of collective risk. While individual users may be willing to "free-ride," collective decision-making processes may be much more effective in curbing overuse. Third, information can play a critical role in the democratization of decision making at the river basin level by training users to make decisions, and by dispelling the widespread distrust that has developed as a result of the traditional patterns of bureaucratic insulation.

Effective participation, however, does not necessarily mean equitable participation. In fact, especially at the local level, there is growing evidence that despite great progress in terms of increased participation, many stakeholders still perceive water management as an exclusionary process. Hence, although there is more room for users and representatives of organized groups to participate, many nonelite groups, such as smaller users or rural agricultural workers, still feel excluded from the process (Ballestero 2004; Taddei 2005). They have shown their discontent with the new model by fiercely resisting several aspects of its implementation—especially water charging—as well as by openly criticizing, and even sabotaging, its implementation (Formiga-Johnsson and Kemper 2005; Taddei 2005).

Similarly, within the negotiated allocation process, Taddei (2005) found that technical language and credentials are often used as an instrument of authority not just by COGERH técnicos but also by elite members over nonelite ones. In this case, technical discourse alienates and overpowers laymen. This contributes to skewing further the decision-making process against nonelite participants (Lemos 2003).

These findings are consistent with data collected by a broad survey of Lower Jaguaribe River Basin Committee members carried out in 2004. Although they believe technical information is useful and helpful to their decision making, they find it is neither widely available nor easily accessible and understandable. They also perceive power within the River Basin Committee as strongly skewed in favor of técnicos over other actors. For example, although 93.1 percent of the members report that technical information makes decision making easier, only 32.1 percent perceive it as accessible and available to all members. Moreover, members surveyed pointed out that the main constraint to the democratization of decision making within the Committee is the disparate level of knowledge between técnicos and general members. This constraint is more important than economic and political power disparities. Indeed, such findings suggest that the persistence of technocratic insulation may be one of the biggest hurdles to overcome in order to increase effective participation in river-basin councils. They also show that despite the best intentions of reform-oriented técnicos, the dominance of technical expertise in water management in Brazil is a difficult pattern to break.

Concerning participation and democratization of decisionmaking, the glass is half-full when we compare the current system to the previously exclusionary, nonparticipatory model. Here the use of COGERH's modeling tools has provided for better-informed decisions and more "efficient" reservoir management. Yet the glass is half-empty because nonelite groups continue to feel alienated from meaningful participation. Such alienation can lead to lack of access to decision making, and ultimately, to lack of access to water. In this case, technical information can further aggravate the situation by limiting access and by providing técnicos with an authoritative voice likely to dominate water allocation negotiations.

Concluding Remarks and Further Research

In the 1990s, Brazil's approach to water reform overhauled the country's old top-down, sectoral system by creating a new set of institutional arrangements that fostered societal participation, integration, and the reframing of water as a common good with economic value. These changes "shook up" water management in Brazil. They allowed for the introduction of unprecedented management schemes with important equity implications. This study focuses on a few of these implications, and in particular, on the role that técnicos and technical knowledge played in the implementation of water reform. In this case, the change from the basic water management paradigm in Brazil provided actors and organizations with greater degrees of freedom both to create new institutions that better fit their water resources and their users management needs, as well as to incorporate new technologies into the decision-making process.

The new system in Ceará adopted a jurisdiction for decision making—the river basin—and created a number of organizations, such as the Users Commissions, which significantly decentralized decision making about water allocation and stimulated user participation. In the context of these new institutions, technical knowledge may have played a critical role in producing better-informed decisions as well as users' heightened perception of efficacy.

However, effects of the use of technoscientific information in the democratization of decision making at the river basin level are more complex. On the one hand, technical information may allow for more participation for water users, especially elites. On the other hand, it may contribute to the continuation of traditional patterns of nonelite exclusion. Further, it may reinforce the dominance of a technical discourse in water management. Advocates for the dominance of technical discourse argue that considering the possibility of excessive and wasteful consumption, there should be limits to users' discretionary powers in the first place. Nonetheless, the Ceará case supports the argument that institutional change alone will not guarantee effective participation of stakeholders in water management. Beyond the creation of formal participatory organizations, the availability and accessibility of knowledge may play a crucial role in improving both policy outcomes (more efficient and equitable allocation of water) and policy process (more transparent, accountable, and democratic decision making).

The role of técnicos and their personal belief systems and worldviews is critical in shaping policy choice and institutional adaptation. As a result, there is evidence—if not of democratic water management—of more equitable, transparent, and accountable systems when compared with the region's previously exclusionary, clientelistic approach to water management. The effective decrease in water resources consumption also indicates progress in intergenerational implications of water management. By the same token, the relative success of negotiated allocation and the role of techno-scientific knowledge use in that success may signal the building of adaptive capacity which will be important to respond to the negative effects of climate variability and change. The failure to consider long-term environmental effects and regulate groundwater use, however, may pose further challenges to future sustainability and the equitable distribution of resources.

The Ceará case offers but a glimpse of the broader implications of water reform for equitable water allocation among different users and generations. More research is needed to understand these processes across different river basins, regions, and countries. So far, our limited evidence suggests there may be significant trade-offs between efficiency and equity. In this context, knowledge can play an important role in illuminating these trade-offs in support of better-informed decision makers. However, since knowledge can also contribute to the persistence of insulated decision-making processes, understanding the kinds of institutional arrangements shaping its use is essential.

Notes

This research is funded by grants from the National Science Foundation (Award #SES 0233961) and the National Oceanographic and Atmospheric Administration (Award #NA03OAR4310010). The study is also part of a broader effort to study water management in Brazil (Projeto Marca D'agua). In collaboration with researchers from Projeto Marca D'agua, a survey of approximately 630 river basin committee members from eighteen different river basins has been carried out. It includes queries on scientific information use and public participation. In addition, ethnographic field research has been carried out in Ceará, Northeast Brazil where we interviewed water managers, river basin committee members and a few managers of the hydropower sector who are active participants in reservoir management. I want to thank Marcelo Flores and Ricardo Gutierrez for their assistance in collecting data in Ceará, and Rebecca Abers, Lori Kumler, and Nate Engle for their suggestions on earlier versions of the manuscript. Finally, I would like to thank the reviewers and editors of this book for their guidance and valuable comments.

1. The four Dublin Principles are: (1) Fresh water is a finite and vulnerable resource, essential to sustain life, development, and the environment; (2) Water development and management should be based on a participatory approach involving users, planners and policymakers at all levels; (3) Women play a central part in the provision, management, and safeguarding of water; (4) Water has an economic value in all its competing uses and should be recognized as an economic good.

2. For an in-depth discussion of pricing schemes in the context of Brazil's water reform see Formiga-Johnsson, Kumler, and Lemos 2007.

3. This finding is consistent with the results of the survey of the Lower Jaguaribe River Basin Committee members who also point to low levels of communication between representatives and represented both before and after User Committee meetings.

References

Abers, Rebecca, and Margaret Keck. 2005. "Águas Turbulentas: Instituições e Práticas Políticas na Reforma do Sistema de Gestão da Água no Brasil" In Marcus André Melo, Cátia Wanderley Lubambo e Denílson Bandeira Coêlho (eds.), Desenho Institucional e Participação Política: Experiências no Brasil Contemporâneo. Rio de Janeiro: Editora Vozes, pp. 155–185.

Adger, W. Neil, Saleemul Huq, Katrina Brown, Declan Conway, and Mike Hulme. 2003. "Adaptation to Climate Change in the Developing World." *Progress in Development Studies* 3:179–195.

ANA—Agencia Nacional de Água. 2002. "A ANA e sua Missão: ser guardiã dos rios." http://www.ana.gov.br/folder/index.htm.

Ballestero, Andrea. 2004. "Institutional Adaptation and Water Reform in Ceará." Master's Thesis. School of Natural Resources and Environment, University of Michigan, Ann Arbor.

Bauer, Carl J. 2004. "Results of Chilean Water Markets: Empirical Research Since 1990." *Water Resources Research* 40 (W09S06):11 pp.

Brannstrom, Christian, James Clarke, and Mariana Newport. 2004. "Civil Society Participation in the Decentralization of Brazil's Water Resources: Assessing participation in Three States." *Singapore Journal of Tropical Geography* 25, no. 3:304–321.

Clemens, Elisabeth S., and James M. Cook. 1999. "Politics and Institutionalism: Explaining Durability and Change." *Annual Review of Sociology* 25:441–66.

Etzioni-Havely, Eva. 1983. *Bureaucracy and Democracy: A Political Dilemma.* Melbourne, Australia: Routledge & Kegan Paul.

Formiga-Johnsson, Rosa Maria, and Karin E. Kemper. 2005. "Institutional and Policy Analysis of River Basin Management in the Jaguaribe River Basin, Ceará, Brazil." *Policy Research Working Paper 3649.* Washington, DC: World Bank.

Formiga-Johnsson, Rosa Maria, Lori Kumler, and Maria Carmen Lemos. 2007. The Politics of Water Pricing in Brazil: Lessons from Paraiba do Sul. *Water Policy,* 9: 87-104.

Garjulli, Rosana. 2001. Oficina temática: Gestão Participativa Dos Recursos Hídricos: Relatório Final. Aracaju, SE: PROAGUA/Agência Nacional de Agua.

Geddes, Barbara. 1990. "Building 'State' Autonomy in Brazil, 1930–1964." *Comparative Politics* 22, no. 2:217–236.

Gleick, Peter H. 2000. The Changing Water Paradigm: A Look At Twenty-First Century Water Resources Development. *Water International* 25, no. 1:127–138.

Healy, Robert G., and William Ascher. 1995. "Knowledge in the Policy Process: Incorporating New Environmental Information in Natural Resources Policymaking." *Policy Sciences* 28, no. 1:1–19.

Kates, Robert W. 2000. "Cautionary Tales: Adaptation and the Global Poor." *Climatic Change* 45:5–17.

Kemper, Karen E., and Douglas Olson. 2001. "Water Pricing: The Dynamics of Institutional Change in Mexico and Ceará, Brazil." In Ariel Dinar (ed.), *The Political Economy of Water Pricing Reforms,* Boulder, CO: Netlibrary, pp. 339–358.

Leach, William D., Neil W. Pelkey, and Paul Sabatier. 2002. "Stakeholder Partnership as Collaborative Policymaking: Evaluation Criteria Applied to Watershed Management in California and Washington." *Journal of Policy Analysis and Management* 21, no.4:645–670.

Lemos, Maria Carmen de M. 1998. "The Politics of Pollution Control in Brazil: State Actors and Social Movements Cleaning up Cubatão." *World Development* 26, no. 1:75–87.

Lemos, Maria Carmen. 2003. "A Tale of Two Policies: the Politics of Seasonal Climate Forecast Use in Ceará, Brazil," *Policy Sciences* 32, no.2: 101–123.

Lemos, Maria Carmen, Timothy Finan, Roger W. Fox, Donald R. Nelson, and Joanna Tucker. 2002. "The Use Of Seasonal Climate Forecasting in Policymaking: Lessons From Northeast Brazil." *Climatic Change* 55, no. 4: 479–507.

Lemos, Maria Carmen, and João Lúcio Farias de Oliveira. 2004. "Can Water Reform Survive Politics? Institutional Change and River Basin Management in Ceará, Northeast Brazil. *World Development* 32, no. 12: 2121–2137.

Lemos, Maria Carmen, and João Lúcio Farias de Oliveira. 2005. "Water Reform Across the State/Society Divide: The Case of Ceará, Brazil." *International Journal of Water Resources Development* 21, no. 1:133–147.

Manor, James. 2005. User Committees: A Potentiality Damaging Second Wave of Decentralization? In Jesse C. Ribot and Anne M. Larson (eds.) *Democratic Decentralization through a Natural Resources Lens,* New York: Routledge, pp. 193–213.

McEwan, Cheryl. 2003. "'Bringing Government to the People': Women, Local Governance and Community Participation in South Africa." *Geoforum* 34, no. 4, November: 469–81.

Mohan, Giles, and Kristian Stokke. 2000. "Participatory Development and Empowerment: The Dangers of Localism." *Third World Quarterly* 21, no. 2, April: 247–68.

Nunes, Edson de Oliveira, and Barbara Geddes (1987). "Dilemmas of State-led Modernization in Brazil." *in* John D. Wirth, Edson de Oliveira Nunes, and Thomas E. Bogenschild, (eds.), *State and Society in Brazil: Continuity and Change.* Boulder, CO: Westview Press.

Rayner, Steve, Denise Lach, Helen Ingram, and Mark Houck. 2002. "Weather Forecasts Are For Wimps: Why Water Resource Managers Don't Use Climate Forecasts." Project Report. National Oceanographic and Atmospheric Administration, Silver Spring, MD: NOAA.

Reis, Elisa P. 1990. "Bureaucrats and Politicians in Current Brazilian Politics." *International Social Science Journal* 42, no. 1:19–29.

Ribot, Jesse C. 2002. *Democratic Decentralization of Natural Resources: Institutionalizing Popular Participation.* Washington, DC: World Resources Institute.

Schmitter, Phillip. 1971. *Interest Conflict and Political Change in Brazil.* Stanford, CA: Stanford University Press.

Smit, Barry, Ian Burton, Richard J. T. Klein, and J. Wandel 2000. "An anatomy of adaptation to climate change and variability." *Climatic Change* 45: 223–251.

Taddei, Renzo. 2005. "Of clouds and streams, prophets and profits: The political semiotics of climate and water in the Brazilian Northeast." Ph.D. dissertation, Graduate School of Arts and Sciences, Columbia University, New York.

Tendler, Judith. 1997. *Good Government in the Tropics.* Baltimore: The John Hopkins University Press.

Wester, Philippus, Douglas J. Merrey, and Marna deLange. 2003. "Boundaries of Consent: Stakeholder Representation in River Basin Management in Mexico and South Africa." *World Development* 31:797–812.

10

Water and Equity in a Changing Climate

Helen Ingram, David Feldman, and John M. Whiteley

The growing recognition of water as a critical problem for the twenty-first century has done little to abate the increasing pressures on this scarce resource. Because water is an essential element in nature and human development, water is touched by everything we do. For instance, humans are changing the climate, and climate change influences water resources. Climate changes are being felt not only in temperature changes, but also in alterations in the amounts and patterns of precipitation and, possibly, in the frequency and intensity of damaging, destructive storms. Further, every attempt made to reduce greenhouse gases by switching to alternative fuels involves new demands on already overstretched water supplies and new threats to water quality. An estimated 40 percent of the world's population lives in areas vulnerable to water scarcity (International Water Management Institute 2007). Moreover, adequate water infrastructure is lacking in many areas. As reported in chapter 1, one billion people worldwide were without adequate domestic water supplies. Water quality continues to worsen in many places at the same time as more than 2 billion people are without sufficient sanitation.

The future challenge is not just to respond to mounting water problems, but to do so in a way that appropriately considers equity. As all the chapters in this volume attest, equity can only be served through processes of decision making that reflect the full range of values with which water is associated. Because many of the issues encountered in the past and likely to be dealt with in the future are complicated by many competing equity claims, no simple utilitarian formula of serving the greatest number of people, or the highest economic value, or the alternative least damaging to the environment or other metric will work. Instead, multiple kinds of rationality and reasoning must be embraced. Diverse constituencies and their legitimate concerns must be accommodated. New life must be given to longstanding tools for incorporating equity into water resources decisions. Greater imagination must be brought to the search for new venues and styles of management.

This chapter will consider a few of the most recent and salient aggravating challenges to water resources that have clear equity implications. The point is that the damaging consequences of large scale water development of dams and diversions are not things that can be assigned to the past. We cannot suppose that we have moved into an era of primarily management problems in which equity may more easily take its rightful place among water goals. Instead, competition among values associated with water supply and management is likely to worsen, and conscientious, forthright efforts must be expended to redesign institutions and policies to ensure that equity issues are addressed–especially in light of likely intensification of conflict among traditional water management goals due to the impacts of a changing climate and pressures of a growing population. Our aim in this chapter is to go beyond raising consciousness of equity issues and critiquing past decision-making processes. We suggest different methods for including and evaluating the full range of values associated with water, as well as the adoption of means, venues, and practices to better incorporate equity into future water decisions. Our goal is to contribute to a future with less legacy of unfairness than plagues water issues today.

Aggravated Problems and the Changing Climate

The consequences of climate change to global water resources will be felt in a myriad of ways, including intensity and frequency of droughts, increased frequency of extreme events such as hurricanes, floods, and wildfires, and the accompanying increased numbers of environmental refugees. The problems of managing water infrastructure including dams and diversions meant to buffer extreme events will become much more difficult. Increased surface temperatures will mean increased evaporation, increased precipitation received as rain rather than snow earlier in the year, and increased water temperature and lowered water quality due to lessened assimilative capacity of warmer waters; and threats to habitat of endangered species. While there may be increased precipitation in some regions of the world, many other regions will become drier, and increases in precipitation may be more than offset by higher evaporation. Sea-level rise, caused by the melting of land ice, will adversely affect coastal areas by increasing flooding, causing salt water intrusion of aquifers, and threatening sources of fresh water (Gleick 2000; Adams and Peck 2006).

The effects to actual localities and particular people of intensified water problems related to climate change will depend very much on the context, including the physical characteristics of the landscape and human resilience and adaptive capacity. It is generally the case, however, that disadvantaged peoples are most vulnerable to climate change– related water resource disruptions. Some analysts have concluded that uneven impacts of climate change will be distributed so that the poorest countries, located at low latitudes, along with island states, will suffer the most. Less developed nations often lack access to technologies that could assist adaptation to climate change. One British charity predicted that there may be up to 200 million climate-change refugees in developing nations by mid-century (Conniff 2007).

Inequities related to climate change may compound those that already exist. The building of water infrastructure worldwide has helped some at the expense of others. According to the World Commission on Dams report, ". . . social groups bearing the social and environmental costs and risks of large dams, especially the poor, vulnerable, and future generations, are often not the same groups that receive the water and electricity services, nor the social and economic benefits of these." (World Commission on Dams 2000) As Vaccaro, Hirt, Mumme, and others in this volume illustrate, politically powerful constituencies have generally been able to shift the burden of water-related risk to others who then bear an inequitable burden. Contemporary water politics is more disruptive and contentious because of the legacy of inequity. The challenge for the future is to find more equitable solutions to water problems, yet most strategies to respond to climate change aggravate equity problems.

The burning of fossil fuels is a major contributor to greenhouse gases. The cooling and emissions scrubbing of electric power generation plants require lots of water. About the same amount of water is withdrawn from water systems for electrical energy as is diverted for irrigated agriculture in the United States, although energy returns a higher percentage of water for reuse. (Hutson et al. 2005). Return flows of water used for cooling usually have higher temperatures and poorer water quality. The deposition of atmospheric pollutants (other than carbon dioxide) from the burning of fossil fuels can degrade water quality in lakes, streams, and rivers. These pollutants include heavy metals and toxic compounds, such as arsenic, selenium, mercury, cadmium, chromium, lead, sulfates, and boron (e.g., Baum 2004).

In an altered climate, these negative side effects will worsen. For instance, during heat waves in the summer of 2007, the Tennessee Valley Authority had to shut down some of its power plants in order not to contribute to already overheated water in the Tennessee River. Increasing reservoir storage as a buffer against drought and a source of hydroelectric energy is not a benign alternative, even when the equity effects of dams are not considered. Large amounts of carbon tied up in trees and other plants are released when reservoirs are flooded. Further, methane from decaying matter that decomposes without oxygen in reservoir bottoms is released when water passes through turbines or when water levels fluctuate exposing such matter on reservoir shores (Graham-Rowe 2005). The alternative of greater dependence on nuclear power plants avoids some of the problems of reservoirs, but at the same time makes large demands on water for cooling and delivers high temperature return flows.

Other "green" energy alternatives to reduce dependence on energy sources that produce large amounts of greenhouse gasses have problematic impacts on the linkage of water and equity. Consider the impacts of greater reliance on biofuels that will both increase competition for scarce water supplies and divert irrigated agriculture from the production of food to the production of oilseeds, such as soybeans, corn, rapeseed, sunflower seed, and sugarcane, among other crops. Focusing first on water availability, parts of China and India have already breached the limit of sustainable water use, without the added strain of trying to grow significant quantities of biofuels. Almost all of India's sugarcane, the country's major ethanol crop, is irrigated, as is about 45 percent of China's top biofuel crop, corn. Both countries depend on increasing the productivity of irrigated agriculture to feed growing populations, but this appears to be inconsistent with plans to increase their reliance on biofuels. In Mexico, one critic of plans to produce ethanol from corn by transnational corporations warns that large-scale production would exacerbate inequities for peasants and rural communities. He further urges those who would pursue this path to emphasize smaller-scale production that will not lead to centralized economic control (Quintana 2007).

According to a presentation by the Stockholm International Water Institute in the summer of 2007, biofuels are water hungry crops that will need about the same amount of water as food crops by 2050, and that rises in the growing of crops for both food and fuel in water-scarce regions does not add up (FO Licht's World Ethanol and Biofuels Report 2007). Clearly, increased demands for irrigation water in China and India will increase pressures to construct more large dams. While such dams might contribute to a more assured water supply, they are likely to continue to have all the equity problems previously identified. Difficulties are predicted even in countries less prone to aridity. In Brazil, for instance, heavy water use to grow sugar cane for biofuels intensifies water shortages and damages river systems (Worldwatch Institute 2007).

Biofuels may also contribute to food insecurity of the poor. Ethanol is predicted to use up half the domestic corn supplies of the United States within a few years (Runge and Senauer 2007). The enormous volume of corn required by the ethanol industry is sending shock waves through the food system. Soaring food prices, driven in part by demands for ethanol made from corn, have slashed the amount of food aid the U.S. government can provide to needy nations. In 2007 the U.S. could buy only half as much food as it did in 2000 (Dugger, 2007). The United States accounts for some 40 percent of the world's corn production, and corn prices have risen sharply. Wheat and rice prices have also surged as more acres are devoted to more profitable corn production. Rising prices of these staples have hurt the poor, and threaten worse effects in the future. In late 2006, the price of tortillas in Mexico, which gets about 80 percent of its corn from the United States, doubled thanks to the rising price of corn.

Mexico has the advantage of petrodollars with which it can, if it chooses, buffer the effects of rising food prices for the poor. Other nations are not so fortunate. For instance, cassava, a basic food for the very poor in Asia has high starch content, which makes it very attractive for the production of ethanol. (Runge and Senauer 2007). While the increasing crop prices could conceivably help poor farmers, land prices have escalated everywhere the demand for biofuel exists, making lands unaffordable to many. An article in *Foreign Affairs* by Runge and Senauer sums up these impacts:

The world's poorest people already spend 50 to 80 percent of their total household income on food. For many who are landless laborers or rural subsistence farmers, large increases in the prices of staple foods will mean malnutrition and hunger. Some of them will tumble over the edge of subsistence to outright starvation, and many more will die from a multitude of hunger-related diseases. (Runge and Senauer 2007)

Not every authoritative commentator on water resources sees the global future of water resources so negatively, even taking into account the additional challenges of global climate change. Some point out that it is more a problem of misdistribution of water resources rather than absolute shortages, and that the crisis is one of governance, not lack of technological, economic and social alternatives to effectively address problems (Rogers 2006).

Acknowledging that water problems are primarily social and political further amplifies the importance of the equity theme of this book. Water is an enormously contested issue. Water-related struggles continue over a number of critical issues, including access to water as a basic human right, the role of markets and private water companies in water delivery, representation and inclusiveness in governance decisions, and what constitutes fairness when equity arguments are made by contending parties to water disputes (Conca 2006).

Norms and Values Associated With Water

The meanings attached to water are fluid and protean. Traditionally, demands on water resources have been classified by the many competing uses: urban water supply, agriculture, industry, energy, recreation, the environment, and so on. Viewed from only this narrow, utilitarian perspective, competition for water resources promises to intensify as demands grow for the production of energy and other uses. Irrigated agriculture is expanding as are domestic water demands of megacities around the increasingly urban world. The lower limits of water supplies and water quality needed to sustain nature and the environment is turning out to be much higher than once imagined. As difficult as it would be to negotiate claims among these users, all are arrayed along the same utilitarian dimension. Additional levels of complexity and conflict arise because there are many other values that are not utilitarian in nature. Meanings of water vary depending on different ways of knowing, varying relationships of water to territory and the state, material and symbolic attachments and other facets or elements (Blatter and Ingram, 2001). Each of the various meanings is the source of claims to legitimacy, voice, and fairness.

While an enormous amount of water scholarship is aimed at establishing recipes, guidelines, best practices, and other templates to impose order, coordination, consistency, and agreement about water, we take a decidedly different view. We agree with the conclusions of Ken Conca (2006) at the end of his book *Governing Water,* and the scientists who issued the Jo'Berg memo (Sachs 2002) that water conflict is inevitable:

> ... communities and individuals bring extraordinarily diverse experiences, interests and worldviews to bear on the global stage. Conflict cannot be dreamed away; on the contrary, conflicts generate the upheavals, alliances and ideologies of that amalgam called global society. There is no universal way of seeing; there are only context bound viewpoints that offer particular perspectives. (p. 67)

Water can be and often is tied to place, or it can be a disembedded symbol or idea. It can be bound up with security. It can be the focal point of community and culture building. It can take on an essentialist character as a gift of nature. Or, it can be an artifact of the human imagination as a matter of taste or lifestyle that may be virtual and artificial. Many of these and other meanings are not commensurable (Blatter, Ingram, and Levesque 2001) and simply cannot be ordered into any kind of priority that simplifies social choices. The multiple and conflicting meanings of water underlie the contemporary conflicts over the role of markets, human rights, geographically based river-basin and watershed institutions, and participation-based processes of inclusive governance. They also help to explain the failure of attempts to construct mutually consistent principles of water ethics.

The UNESCO formed World Commission on Ethics of Scientific Knowledge and Technology is a case in point. Searching for universals, the commission offered in 2002 six water principles:

1. Principle of human dignity—water as a basic human right;
2. Principle of participation—focus on citizen participation in decision making;
3. Principle of solidarity—we all rely on the continued health of our ecosystems and are linked through our upstream and downstream dependency on these systems;
4. Principle of human equality—incorporating the values of justice and equity;
5. Principle of common good—water as a common good and essential to the realization of full human potential and dignity; and
6. Principle of stewardship—moving toward a sustainable ethic and finding a balance between using, changing, and preserving our land and water resources (Selbourne 2000).

While such a list of principles contributes usefully to discourse and discussion, they are not a practical list of rules that can be applied to real-world situations, as this volume illustrates. While access to sufficient water to sustain life probably should be classified as a basic human right, water still has to be delivered from one place to another through infrastructure that has economic and social costs.

As chapters 5 and 6 illustrate, there are winners and losers involved in trying to deliver water to serve the exploding population along the U.S.-Mexico border, and certainly guaranteeing everyone access to water does little to secure sustainability or stewardship. Public participation can empower previously disadvantaged participants in water conflicts as is illustrated in chapter 2, but such participation in many other circumstances may do very little to secure minority rights including access to water or to guarantee the rights of future generations. Linking upstream and downstream communities into a systemic perspective can be quite problematic, as shown in chapter 3.

The last of the six principles most clearly reveals the fundamental flaw of the attempt to prescribe universals. Finding a balance is, after all, the problem. Declaring that balance should be found is hardly a recipe for doing it. Since water conflict is inevitable, and cannot be avoided through any application of universal principles, attention needs to shift to processes, tools, and venues, or institutions that can deal with conflict while attending to equity.

Normative and Communicative Reasoning: Means and Tools for Incorporating Equity into Resources Decisions

Several contributors to this volume have noted that policies designed to pursue "the greatest good for the greatest number," or "maximizing social utility," have been defended—in some cases—as means of generating "fair" or equitable water policy. Chapter 3, on the storm water drainage policies in California, for example, employs utilitarianism, particularly as it relates to the choice of actions leading to the best consequences, and procedures affording everyone an equal opportunity for participation. The analysis in this chapter of the misdistribution of benefits and costs among affected parties and the absence of open and equitable procedures illustrates the usefulness of utilitarian thought. As other contributors have also demonstrated, however, such utilitarian approaches have a number of long-standing deficiencies, among them is reliance on individual self-interest as fundamental. The present discussion is an effort to identify sound ethical alternatives to utilitarianism for managing water resources, the adoption of which could be used to bring about the goals of equity addressed in this volume. Before discussing these alternatives, however, we must begin by acknowledging the fact that implementing any alternative will be difficult for three reasons.

First, there is a long-standing bias, in Western societies in particular (reflected in the politics of both developed nations and in the corporations, NGOs, and international aid organizations they finance, support, and house), toward defending policy decisions on the basis of rational self-interest. This bias inhibits public interest-based solutions to environmental and resource issues because it assumes that individual economic gain is itself a virtue. It rejects civic republicanism, or the conviction that citizens have an obligation to serve the needs of the larger community (Bellah 1996). This is a problem explicitly acknowledged in particular in chapter 2 on the San Luis Valley and chapter 6 on the Pacific Northwest.

Second, each ethical alternative discussed here has significant shortcomings when taken (or applied) alone. This is evidenced by the fact that all three have been, at least in part, utilized throughout history as justifications for environmental policy decisions. In order to be effective, therefore, these alternatives *must* be implemented in ways that embrace the needs of watersheds *and* the demands of democratic theory, or they will fail. Such a "social ecology" approach is consistent with the lessons predicated in chapters 5, 7, and 8.

Finally, it is difficult, at best, to achieve clear and transparent policy objectives that are ethical as well as practical through a political process characterized by interest group negotiation, bargaining, and brokered compromise. Ethical discourse is concerned with defining fundamental issues of right and wrong. Politics, on the other hand, as the late political scientist Harold D. Lasswell reminded us, is the study of "who gets what, when, and how" (Lasswell 1958). Figuring out how to use power to achieve one's goals is at the heart of politics. Determining how to exercise that power justly is a secondary concern—or, worse, is ignored entirely. For these reasons, there are no ethical panaceas for achieving a just environmental policy. As chapters 4, 5, and 9, in particular, remind us: where national or regional parties to water use have divergent power, achieving equity in allocation and distribution, and fairly adjudicating disputes over water rights, is nearly always intractable.

Thus, our challenge is to identify approaches to the implementation of ethical ideals that can temper the struggle for justice with appropriate indignation toward pure self-interest on the one hand, and due regard for the inevitability of the struggle for power—endemic in all polities—on the other.[1] Furthermore, and consistent with the arguments in several chapters in this volume, while utilitarianism may promote efficiency, the latter is, at best, an instrumental *ethical* criterion whose purpose is usually thought to be the achievement of some higher goal. It is by no means the only basis for defining a rational choice. As Charles Anderson notes, other principles, including a policy's amenability to ranking conflicting social priorities, and its ability to mitigate risks imposed on less fortunate populations—or vulnerable resources (e.g., see chapters 3 and 9) are also important considerations in determining what makes a decision rational (Anderson 1979, 711–723). Economically efficient water policies have often imposed adverse ecological and social risks on some, as these chapters show. Clearly, an equitable water policy must embrace a broad conception of individual and collective welfare.

This is one reason that equity is essential in water policy: utilitarianism cannot serve the common good unless some allowance is made for how the benefits of natural resources management are allocated. As Ralph Ellis notes:

[No] matter how the principle of utility maximization is interpreted, dressed up, or socio-politically contextualized, there is no way to erase the fundamental conflict between the value of maximizing human well-being on the one hand (however 'well-being' is conceived), and on the other hand, the fairness or justice of the way this human well-being is distributed among the population, either in the form of desirable benefits and situations for people or in the form of opportunities to achieve (them). (Ellis 1998, 84)

Even among advocates of the supremacy of efficiency, the importance of distributive justice in natural resources policy has been an important determinant of schemes for regional resource development, such as the Tennessee Valley Authority. It has also been one impulse giving rise to a number of federal programs and projects to develop western water resources, justifying them not only as compensation for the hardship endured in western migration, but as a means of developing resources whose benefits would extend to easterners as well, as pointed out in chapter 6.

We now discuss three means for incorporating equity: covenants, categorical imperatives, and environmental stewardship. Table 10.1 summarizes the major features of these approaches.

Covenants as an Equity Alternative for Water Resources Management

The concept of a covenant as a means of defining our responsibilities for managing natural resources, including water, and the benefits received in managing them well began in ancient Israel about four thousand years ago—although archaeologists note that covenantal language was also used throughout the ancient Near East even earlier, having been employed by officials in ancient Assyria and Sumeria, for example (Bright 2000). While often confused with contracts—"tit-for-tat" legal agreements—covenants are based on considerably more than mere reciprocity.

Specifically, covenants are predicated on three principles. First, by agreeing to specific laws and responsibilities regarding the management of natural resources, people within a geographical region achieve a dominion, or right to rule, over nature.[2] It is not enough for an individual to agree to these stipulations; an entire society and culture must submit to this framework. Only a few individuals need to break this covenant for many to suffer the consequences of doing so; the entire society bears collective responsibility for the protection of natural resources. Second, covenants, unlike mere contracts, are deemed permanent. Third, covenants often derive their authority from a deity. This places them on a higher moral plateau than mere contracts. Consequently, the dominion granted to mankind for covenantal enforcement explicitly embraces *all* forms of life. In effect, a covenant differs from a contract in that obedience is predicated not on prudential, human-centered interest, but on reverence and respect for a higher and more noble interest—God's desire that those created in his image care for and protect all of his creation, not just themselves.

Today, it might seem strange to use the term "covenants" in relation to equity in water resources policy. In fact, however, covenantal language implicitly pervades past and current water resource management methods, and covenantal lessons remain relevant to contemporary water policy. A good, albeit imperfect, example of the covenantal principle is the concept of a river basin compact. As the historian Joe Gelt argues, the 1922 Colorado River Compact was much more than a contractual agreement among seven states. It was the keystone to the so-called "Law of the

Table 10.1
Three means to incorporate equity in water resources decisions—comparative criteria

Ethical Approach	How do they address traditional controversies?	Advantages	Disadvantages/drawbacks
Covenants	(1) Neutral regarding structural/nonstructural approaches—decide on basis of what is most "righteous" solution; (2) favors comprehensive planning ; (3) encourage public awareness and participation in decision making and a public process of responsibility.	(1) Focuses on mutuality of obligations; (2) Offers guidance for specific structure of political frameworks such as river basin compacts.	(1) Evidence of actual historical covenants are difficult to identify—they are more of a logical than an empirical artifact; (2) may be difficult to change a covenant once it is consummated—thus, may be problematic for democracy; (3) historically, obligations to nature are unclear—is it subjugation of nature, or protection of resources?
Categorical imperatives	(1) Structural/nonstructural debate boils down to impact on fulfilling promises, commitments, and other intrinsic moral properties; (2) comprehensive planning important to ensure fulfillment of obligations—and to ensure they can be enforced across generations;	(1) Permits management approach to be modified in face of new knowledge and information; (2) provides early warning of problems that inhibit fair/honest/judicious decisions (e.g., irreversible impacts)	(1) Does not adequately embrace neither non-renewable nor endangered resources—problematic for watersheds; (2) Does not provide a clear means of determining whether or when an intrinsic promise or other commitment outweighs the merits of a teleological

	(3) participation required to ensure that promises/other obligations are freely and explicitly made.	argument (e.g., a more traditional "benefits" vs. "costs" approach—for this reason problematic for democracy.
Stewardship	(1) Weight of evidence supports nonstructural approaches to minimize adverse impacts and to be faithful to protection of creation; (2) planning should be done in an organized, comprehensive way to protect creation; (3) public participation must include broad range of issues/concerns including nature—not just people.	(1) Tries to balance human and ecological needs—rejects the notion of human "dominion" over nature; (2) does not supplant or replace other approaches—is really a supplement to them.	(1) Reasons for caring for nature somewhat ambiguous—is it a "moral imperative" enforceable by public policy, or an act of unselfish love which requires a change of attitude? (2) requires a deep sense of humility toward nature.

River." Advocates of the "Law of the River" wanted to avoid, as much as possible, both protracted litigation and federal intrusion in managing the region's scarce water. These advocates agreed to apportion the waters of the river "in perpetuity" between upper and lower basins (states would later work out their percentage allocations).

That the Compact did not adequately embrace environmental issues and Indian water rights is, as Gelt notes, a function of the era in which it was consummated, not an inherent defect of compacts. At the time, ethics were less a concern in concluding an agreement than political expedience. As Gelt notes:

An environmental ethic arises as a force in contemporary life through a somewhat different historical process than, say, water marketing and to some extent Indian water rights. Espousing an environmental ethic involves a shift in thinking, a reorientation of values, away from the human-centered and toward acknowledging an obligation to the natural world. *Development,* however, was the overriding concern of the 1922 compact. Its intent was 'to secure the expeditious agricultural and industrial development of the Colorado Basin, the storage of its waters, and the protection of life and property from floods.' Establishing Colorado River rights was a prerequisite to building flood control and storage projects, to better manage the river to serve human needs. This boosted states' potential to grow and develop. (Gelt 1997, 1)

A better example of a contemporary covenant that protects human and nonhuman equity is provided in chapter 2 on the San Luis Valley. In this case, the Great Sand Dunes National Monument, augmented by the water rights secured through the collaborative efforts of the San Juan Valley residents and The Nature Conservancy, reflected a promise in perpetuity, not just to protect the natural features of the park, but also the history, culture, and entire way of life of the San Juan Valley. While the author writes in terms of moral economy, not covenant, and stresses analytical principles, not means, the history he recounts shows that the mechanism chosen to embody equity is not unlike the Colorado River Compact. Indeed, critics of national parks have often reflected on the way in which the designation of "national park" locks in existent uses and nonuses of natural resources such as water.

As it pertains to the management of natural resources, the Old Testament contains several illustrative examples of how a covenantal relationship functions, and its political consequences. In the Book of Genesis, the story of Noah's Ark obviously has water as a central focus. As punishment for mankind's becoming "corrupt and full of violence," Yahweh brought about the Great Flood to cleanse the earth of evil. What is often forgotten about this story, however, is that once punishment was meted out, a restored humanity (in particular, those found "righteous" and capable of exercising moral responsibility) was given the opportunity to submit to a new covenant. This covenant promised abundant prosperity and protection from future calamity in exchange for agreeing to obey God and to care for nature—benefits enumerated quite clearly in Genesis, at the foundation of the Hebrew nation, and in Deuteronomy, when Moses "read" the law to Israel.[3] This covenant embraces *all* forms of life, and is based on submission to one another's interests and to God. Clearly, this entails a far broader conception of cooperation—say, as applied to the operation of a river basin authority or other political arrangement for governing water resources—than that characterizing most legal agreements.

A second example of the benefits of a covenantal relationship is from the Book of Ezekiel, in which the prophet Ezekiel shares his vision of a restored nation that one day will again find life in Canaan with the Hebrews living in captivity in Babylon:

Swarms of living creatures will live wherever the river flows. There will be large numbers of fish, because the river flows there. . . . The fish will be of many kinds. . . . Fruit trees of all kinds will grow on both banks. Their leaves will not wither, nor will their fruit fail . . . because the water flows to them. (Ezekiel 47: 8–12)

The significance of this story lies in its implicit recognition that both biodiversity and a sustainable environment are characteristic of freely flowing, unfettered waters. Moreover, such an environment is a gift in perpetuity for nations that submit to God's authority.

In more modern times, covenantal approaches to managing natural resources have been characterized by considerable variability, ranging from Thomas Hobbes' notion of social contract, which directly influenced the emergence of the liberal state and its emphasis on protecting individual rights, and was based largely on prudential self-interest,[4] to John Rawls' *A Theory of Justice*, which offers a covenantal approach to political decisions (Rawls 1971). Rawls would endorse the principle that rational, moral individuals agree on the allocation of water and other natural resources through agreeing upon a set of formal principles for their regional management. These principles for water use are guided by the premises that water is an essential good, and that each person's use of it has some impact on the quantity and quality available for the next user. Pareto optimality is employed to assure that no change in use or allocation would be permitted unless the change makes others better off without making anyone worse off—an ideal for achieving distributive equity also discussed in chapter 1.

Inequalities can be justified *if* they result in an increase in efficiency that makes everyone better off. For example, a new irrigation or public water supply project could be defensible, even if the costs of building it are borne by everyone in a society (and not just the project's direct users), if it leads to the production of more food and fiber (which, presumably, would benefit everyone), and if there are no identifiable alternative means of promoting these same ends.

Finally, the Rawlsian approach is consistent with the premises of democratic theory. Like the Hebrew conception of covenant, it obligates citizens to reject narrow self-interest in favor of actions that make us better citizens. This view concedes we are all finite, fallible creatures and that, before we can make decisions over water management that promote a tangible, material end, we must first agree on procedures that define the "rightness" of an allocation system—the kind of system that would be just.

Another important issue related to covenants is the fact that, in the New Testament, the foundation of a new covenant between God and humans was viewed by early Christians, as by the Hebrews, as embracing nature *and* humanity. A major difference, however, is the subtle—but important—movement away from dominion to stewardship. The Apostle Paul, for example, viewed the freeing of man from the bondage of sin as part of a larger liberation of nature from exploitation and "decay."[5] This new covenant obliges us to care lovingly for the resources provided by God, not to dominate or to exploit them—a theme, as noted in chapter 2, which transcends a number of cultures.

While covenants focus on the mutuality of civic obligations and provide guidance useful for the structure of political frameworks for management, they harbor three potential drawbacks. First, evidence of actual historical covenants is difficult to identify. In fact, covenants are more logical than empirical artifacts; they illustrate how political and legal obligations should come about in the absence of authoritative institutions (Rawls 1971).

Second, since a covenant is rooted partly on the principle of perpetuity, it is difficult to change once consummated, even if conditions for change are warranted based on new knowledge about the actual distribution and availability of a resource, or because of demographic or other social change. This is potentially problematic for effective water management.

As several chapters in this volume show, much of what we now know about the behavior of watersheds, and how agreements to manage them can cause harm, has only been learned in the past three or four decades through the conscientious application of dedicated watershed science (see, for example, chapters 2 and 6. One practical criticism of the Rawlsian formulation of the social contract is that Rawls himself claims that, in attempting to achieve justice, the actual economic wealth of a nation is unimportant in defining the potential for justice (e.g., Caney 2001, 984–986). In reality, as noted in chapters 5, 8, and 9, the overall prosperity of a country can be extremely important in determining whether or not parties to an agreement are comfortable negotiating modalities for governance that may lead to some redistribution of both wealth and political authority.

A final problem with covenants is that our resulting obligations to nature remain unclear: are we permitted to subjugate other forms of life within watersheds for our own benefit, or must we protect nature for its own sake? Recent investigations from a Judeo-Christian perspective suggest that the problem can be resolved if we understand that the dualistic division between humans and nature—a distinction derived from the hierarchical "order of creation" concept, in which humans are viewed as separate from the rest of creation—is incorrect. A preferred alternative, it has been argued, is the notion of "integrity of creation." Stated simply, this means recognizing creation as a holistic unit and emphasizing the communal relationship between it and a creator (McFague 1993).

Categorical Imperatives, Environmental Ethics, and Water Management

Immanuel Kant, the German philosopher most closely identified with the concept of the "categorical imperative," was interested mainly in moral philosophy and ethical relationships among people. Nevertheless, because Kant's ideas were partly a response to the dominant position held by utilitarian views, his ideas have important implications for natural resources management. The categorical imperative, says Kant, is a:

command . . . which present(s) an action as of itself objectively necessary, without regard to any other end. . . . It is (to) act only according to that maxim by which you can at the same time will that it should become a universal law. (Kant 1975)

A moral decision should not aggrandize our own happiness, Kant believed, but be generalizable to all who face a comparable choice in a similar situation. This core ideal is not radically divergent from a covenantal approach. In practical terms, Kant was really asking us to "do unto others as we would have them do to us."[6] How does this apply to natural resources management, and especially to river basin policy? In effect, while some view the ethical basis for managing resources as a matter of maximizing social utility, others view it deontologically; that is, what is good or right is determined by the intrinsic properties of an action, such as promises, commitments, and obligations.

Several chapters in this volume have stressed the importance of promises and prior commitments—animated and reinforced by a strong sense of territorially based community as a means to ensure equity. Other examples abound. Observers of prior appropriation law in the Western United States, for example, note that the "first in time, first in right" principle in effect asserts a sort of preemptory principle: historical commitments should take precedence over other demands. Thus, offstream uses that are deemed reasonable and beneficial take priority over instream needs (Pearson 2000, 24–28). As noted in this volume's introduction, such an approach embraces both compensatory and, to a degree, covenantal principles.

Historically, under appropriation law, "instream uses (were) considered inherently wasteful since they require(d) water to remain in place." The result of this legal interpretation biased the concept of "beneficial" to mean offstream, consumptive uses that divert water from a stream "in order to perfect an appropriation (Butler 1990)." This conception of beneficial use has undergone change in the past few decades, with courts more recently deciding in favor of fishing and recreation as beneficial uses, and also making greater allowance for Native American reserved water rights and minimum stream flows (especially in waters coursing through public lands), even if some senior prior appropriators' rights are impaired as a consequence (Burton 1991; Laitos and Tomain 1992; Landry 2000; Restoring the Waters 1997).

A practical issue that arises in applying categorical imperatives to the problem of water and equity is: what happens if the conditions existing when a freely made commitment or promise was initially made undergo change, as could occur through urbanization, climatic change, or even political revolution? Moreover, how does a promise made to one generation inhibit opportunities for future ones? The literature suggests three possible scenarios.

The first approach is a *restricted-choice argument,* which contends that all of us, regardless of the generation into which we are born, have—at birth—free will and the capacity for independent choice. Any environmental decision made by people in any society that precludes or denies this capacity for exercising free will is morally wrong because it denies presenting future generations the ability to select the benefits they feel are most important, and denies future generations the opportunity to freely change decisions in ways that best fit their choices (Beatley 1994).

Environmental examples of restricted choices include such activities as clear-cutting old-growth forests, filling in wetlands, and building large dams. In effect, such actions restrict freedom of choice by: potentially denying a whole range of future benefits without the consent of those affected by such decisions (e.g., receiving the flood mitigation benefits of a swamp; experiencing the beauty and tangible environmental benefits of an old-growth forest; being able to fish for salmon in a free-flowing stream); and by generating irreversible impacts. For example, tearing down a dam may not automatically restore riparian stream conditions. In effect, the reversibility of an action creates a sort of ratcheted obligation. The more reversible the action, the less risk we impose on the future; thus, the less obligated we are to consider the long-term impacts of our decisions. The less reversible an action is, the greater the risk it poses to future generations. It is inherently wrong to deny to future generations the capacity to exercise a choice that we would have expected past generations to extend to us (Rolston 1988).

As noted in the chapters on Bolivia, northern Spain, and Brazil, there is a host of reversibility problems related to the management of water resources and equity for which too little regard is paid traditionally by policymakers, including: promises that water will always be affordable and fairly priced, assurances that traditional populations' rights of access will be protected, commitments regarding the protection of traditional economic lifestyles and community identities, and the assurance that traditional quantities (and high quality) of water will be provided to people regardless of changing economic or ecological conditions.

A second approach to the categorical imperative problem is the *labor value argument,* also relevant to the welfare of future generations. It differs from the first approach by emphasizing the manner in which environmental value is initially determined. Following the arguments of classical liberals such as John Locke, this approach contends that the value of land, water, and other natural resources is determined by the amount of effort we infuse into their development and management. When we fail to leave the land and water resources of our world in a condition in which they can be at least at the same level of productive benefit for future generations as for us, we not only deny future generations the direct benefits of the resources, such as consumption and use, but we leave them less well off in other, indirect ways—for example, through reduced productivity of farmland, reduced fishery yields, and even a lower tax base and foundation for economic development (Kavka 1978). Chapters 2, 5, and 8, in particular, note the importance of incorporating the welfare of future generations in decisional frameworks predicated on ensuring equity and fairness.

An alternative definition of "sustainable development" could be managing renewable resources in such a way as to support the economic and environmental needs of future as well as present generations. In effect, each generation is obliged to make certain assumptions about the needs of subsequent generations—including how much food will be required, as well as how much land and water, and in what kind of condition, is required to sustain them (Beatley 1994). We also are required to embrace ways of making decisions that ensure these considerations will be met.

Clearly, while categorical imperatives articulate intrinsic principles to guide behavior, they also impose on us the further obligation to amend or modify how these principles are applied as science and other sources of knowledge modify our understanding of the environment—to adopt policies that will conform to the paradigm of adaptive management implied by a number of chapters throughout this volume, and by a number of other recent works on water policy and its reform—and to do so through institutions that encourage wide dissemination of water knowledge (e.g., Moody 1997, 7; Feldman 2007; Lee 1993; also, see chapter 9 in this volume). A more problematic challenge to categorical imperatives for managing water is afforded by climate change. This is an issue that plays out in various ways, as noted at the beginning of this chapter. The balance of evidence suggests not only that climate change is occurring, but also that previous climate changes may have dramatically altered the flow of many rivers and streams worldwide (Meredith 2001; IPCC 2007 a, b). Both facts exacerbate the uncertainty of being able to adhere to a categorical imperative. There are two reasons for this. First, when agreements to allocate water—especially in large river basins—are consummated by compact (the Colorado River basin is a good example), allocation of flow, particularly to various portions of a basin, are predicated on an assumed annual flow that may not hold true today, or may (again, using the Colorado River as an example) be atypical of the mean annual flow of the river, because precipitation and flow were anomalously high during the period prior to the Compact negotiations. Thus, the promise of a given allocation may be based on a faulty premise, and political jurisdictions may use more water than is generally available.

Assumptions may later be interpreted very differently by parties to the agreement. In the case of the 1944 treaty between the United States and Mexico, designed to apportion specific flows of Colorado River water to Mexico, nothing specific was said about water quality. As explained in chapter 5, Mexico assumed that it was due water of good quality while the United States asserted it had no such obligation.

A second challenge is that changes in a regional economy exacerbate the difficulties of providing agreed-on allocations of water, especially given the likelihood of diminished future supplies due to climate change (Meredith 2001, 13). To some degree, this is a lesson embedded in the chapters regarding Mexico and the United States, and Northern Spain. Finally, and especially relevant for the contributions to this volume, equity principles based on categorical imperatives imply a high level of active participation by protagonists; otherwise there is no way to guarantee that promises and commitments are freely made and explicitly articulated. It is interesting that venues for more participatory decision making is built into the design of more recent international water institutions like the La Paz Agreement and the post-NAFTA border commissions. As observed in chapter 5, however, participation cannot occur without sufficient financial and human resources.

Environmental Stewardship and Environmental Ethics: Ruling Like a Servant

Perhaps the simplest way to conceive the meaning of stewardship is via the aphorism: "we have not inherited the environment from our grandparents, we are only borrowing it from our grandchildren." Not unlike the two previous approaches, stewardship ethics are based on the premise that we are obliged to care for creation and to concern ourselves with what anyone—regardless of generation—must do to ensure that creation is sustained. This obligation is rooted in a humble anthropocentrism that, instead of putting mankind at the center of the world, asserts that all species have inherent value and that all individuals have moral standing. Because humans are made in God's image, humankind is both within, as well as above, the broader ecosystem (Grizzle and Barrett 1996).

While the exact meaning of stewardship is subject to debate, at its center is the notion that humans are responsible for caring for the natural environment (in the Judeo-Christian and other faith traditions, for "creation"). Not only is nature unique, but also sacred (Fowler 1995). In the Judeo-Christian tradition particularly, stewardship has become linked to the role of sin—operationally defined as the foundation of failed stewardship or the failure to treat creation as a holistic unit by violating its integrity. The practical manifestation of this is the relationship between rampant environmental degradation and spiritual decline. Calvin DeWitt suggests that this spiritual decline is manifested in our often incomplete understanding of dominion over nature, which does not commend us to exploit land and water resources, but to "till and keep" the earth "with full integrity and wholeness," thus protecting the soil, air, and plant and animal communities (DeWitt 2002; also, Genesis 1:28).

While contemporary stewardship ethics owe a debt to the Judeo-Christian tradition (Fowler 1995, 158), it is also indebted to the more pragmatically derived ecological views that emerged in the late nineteenth and early twentieth centuries—most notably through the writings of Gifford Pinchot, the first director of the U.S. Forest Service; John Wesley Powell, the first director of what would become the U.S. Geological Survey (and the Colorado's first river runner); and later, the naturalist and forester Aldo Leopold.[7] These advocates of the "progressive conservationist" tradition understood humanity's obligation to care for nature to be rooted in the unique stature held by people—as creatures of reason with a capacity to serve as caretakers and guardians of natural resources. Leopold added another stricture: in caring for natural resources, we care for our own welfare and are cognizant of nature's limits. As he stated it, "A thing is right if . . . it preserves the integrity, stability, and beauty of the biosphere" (Leopold 1949).

As previously noted, stewardship approaches to river basin management do not obviate or supplant the other two approaches. One should (perhaps must) adopt a sort of covenantal understanding of one's obligations to nature in order to practice stewardship. Likewise, there is intrinsic value—a categorical imperative in the stewardship stricture to "care for creation" unselfishly, as a servant-ruler.

The Northwest Power Planning Council, discussed in chapter 6, is a fitting example of river management that manifests the goal, at least, of stewardship—the need to balance creation and human interests. The council's purpose is to elevate the concept of river basin management to a higher level—to cement a new set of relations among water users in the Pacific Northwest in an effort to restore salmon spawning runs on the Columbia and Snake rivers. What makes the Planning and Conservation Act unique is that it makes salmon a "coequal partner" with hydroelectric power in the operation of the Columbia Basin's more than 150 dams, and also makes conscious the ethical implications of water use in basin-wide decisions (Blumm 1998; Volkman 1996). Clearly, NPPC is a major departure from previous federal water supply planning efforts in the United States that also illustrates the practical ambiguities in stewardship approaches to river basin management. While these approaches suggest that caring for the environment is a moral imperative, they also imply "intense, interactive love relationships" between humans and creation (Fowler 1995, 82).

Finally, by rejecting the dominionism characteristic of earlier approaches to environmental ethics, including early covenantal approaches, stewardship forces us to eradicate the dualistic split between humans and nature, a distinction originally derived from the hierarchical "order of creation" concept, in which humans are viewed as separate from the rest of creation (McFague 1993). This alternative to dualism focuses instead on the "integrity of creation." Stated simply, this means recognizing creation as a holistic unit, with an emphasis on the communal relationship between people and the rest of creation—a concept derived in part from ecological principles of complex ecosystems and the interrelatedness of their parts. Each part has a function, and the whole system is changed when any part is changed. In various ways, the chapters by Mumme—stressing a broader view of ecological security that transcends nation-states, and Arnold—emphasizing how sustainability is an inclusive concept that must embrace the fate of people as well as endangered resources in a region—are excellent examples of cases that lend themselves to the application of stewardship approaches to equity as outlined here.

Watershed-based Venues and Social Learning

As the contributors to this volume have shown, an *effective* approach to the problem of equity in managing water resources is one that, when implemented, embraces the needs of watersheds *and* the demands of democratic theory. This is a difficult policy challenge, in part, because few philosophical approaches to equity are developed with the specific purpose of being disseminated in the form of laws or public policies. In addition, none of the three approaches we have discussed can embrace every contingency discussed in the cases in this volume.

Implicit in our discussion is the notion that adaptive management is not only a scientific concept; one based, in other words, on the principle that decisions regarding water can be made and modified on the basis of the best scientific information available—but is an ethical concept that embraces equity. In effect, a water management organization may be considered adaptively managed if it has a well-developed feedback process that permits institutional change in light of new information; permits participants–including the broad public—to find common ground in identifying solutions to water problems; and, encourages social learning—the ability to process new information about the natural and social environment and then implement this information. Social learning not only allows for rapid adaptation to new circumstances, it also permits decisions to be made and modified fairly and equitably by transforming policies and programs and encouraging reform of the organizations that manage water. This is an important lesson of this volume. How these objectives can be achieved requires some consideration of the meaning and various dimensions of social learning.

Adaptive Management and Social Learning

Adaptable regimes develop mechanisms for collaborating across diverse groups, learn from mistakes by instituting elaborate feedback mechanisms and participatory innovations, and can reverse flawed decisions, in part because they choose nonstructural solutions. Moreover, adaptively managed organizations recognize that most problems are too complex and ill-structured for single-discipline answers. Thus, they respond to problems by aiming for comprehensive solutions that avoid incompatible objectives (Ellis 1998; Blatter and Ingram 2001; Ingram 2001).

To be more specific, a social learning approach affords public officials, NGOs, and the public the opportunity to adapt public policies as new conditions and other compelling evidence warrants. Normatively, this approach compels us to: (1) admit mistakes; (2) monitor and measure change; and (3) adopt midcourse corrections to policies (further implying that water management decisions should be tentative and reversible). Let us explore these assumptions in turn.

Learning from Past Mistakes

Learning from previous mistakes regarding equity in water policy requires that we be willing to evaluate previous decisions in light of our failures to define what we want and agree on basic goals. If we could do this, we could then learn from our errors and avoid them in the future. In too many instances, failures in water policy under the utilitarian paradigm came about because policymakers sought to please every constituency without facing the incompatibility of policy goals—to say nothing of the narrow basis on which these goals were often predicated.

Learning from mistakes requires adopting a decision-making process that permits the airing of the widest possible sets of values and preferences; the acceptance of ground rules for what will be accepted as credible evidence to support these values; a willingness to make decisions tentatively, with allowance for their reversal if they are found faulty; modification if changing conditions warrant doing so; and avoidance, under any circumstances, of severe impacts. As several chapters in this volume have noted, this requires an openness to adopt watershed and river basin management plans that ensure protection and recovery of species threatened by ecologically damaging human actions, and a kind of humble anthropocentric view characteristic of stewardship to grant moral standing to nature as well as humans.

Monitoring and Measuring Change

Social learning within a watershed venue also requires good information and data. But what makes data "good"? This is a question alluded to by several authors—particularly Lemos, Hirt, and Arnold—whose chapters ponder the problem of ensuring that data is useful as well as valid for watershed decision making. We commonly accept that good data is collected and analyzed dispassionately and without regard to political ends. It also avoids a predisposition for selective promotion of certain claims that serve to advance the cause of one group at the expense of another— a practice that has become commonplace in U.S. environmental policy (Kantrowitz 1993, 101). This so-called advocacy science, which has become a common feature of environmental decision making generally, and water policy in particular, inhibits adaptive management (Ingram 2001). How can change be measured and monitored in an ethical manner? The categorical imperative approach provides one avenue: in essence, adopting no policy outcome for water management that one would not accept if another party adopted it because of fear of the consequences to oneself.

As several cases in this volume note, however, politically weak protagonists often cannot avail themselves of good research findings, may have political difficulty confronting detractors, and are faced with difficulties in resisting pressures for rapid decisions rather than the sort of longer-term scientific experimentation required for ensuring more environmentally benign policy choices. Overcoming this problem requires decision-making venues that compel those in power to submit themselves to the scrutiny of those affected by their power and authority—and to make accessible the dissemination of scientific information to various audiences, not only experts.

Making Midcourse Corrections

There is wide agreement among students of adaptive management (e.g., Hartig et al.; 1992; Landre and Knuth 1993; Cortner and Moote 1994) that approaches to water management that are most advantageous to averting conflict and ensuring fairness are ones that permit protagonists to evaluate systematically structural as well as nonstructural alternatives before making rational and fair policy choices. Such an aspiration is found in the executive summary of the somewhat less than enthusiastically received Western Water Policy Review Advisory Commission's final report. Congress charged the commission with formulating recommendations on the federal government's role in managing the region's water. Its charge fits well with the notion of being able to make midcourse corrections to policy. As stated in its preamble:

Part of the impetus for our Commission's formation was the Congress's finding that current federal water policy suffers from unclear and conflicting goals implemented by a maze of agencies and programs. . . . Lack of policy clarity and coordination resulting in gridlock . . . cannot be resolved piecemeal but, rather, must be addressed by fundamental changes in institutional structure and government process. . . . the geographic, hydrologic, ecologic, social, and economic diversity of the West will require regionally and locally tailored solutions. (Water in the West 1998, xiii)

It is undeniable that many of the problems giving rise to inequities came about because of large-scale engineered efforts to divert, appropriate, and control the flows of rivers and streams. As the chapters by Kaminiecki and Below, Hirt, and others suggest, when we engage in wide-scale re-engineering efforts in an attempt to better manage water supply, or better configure the conditions that enhance its quality, unanticipated consequences often arise that force us to rethink the wisdom of what we have done. Unfortunately, while we weigh those consequences, harm may already have been done to various groups.

Watersheds as Inclusive Venues

While claims regarding the desirability of integrated management schemes based on watersheds make intuitive sense, political scientists and others who study water resources recognize that they face a difficult challenge in implementing them: how to reconcile the roles of science and public values in decision making. While these roles need not necessarily be at odds, how to reconcile them is by no means obvious. An emerging consensus is that reconciliation must begin with an astute public being scientifically equipped to identify salient problems within a watershed they want assessed (Landre and Knuth 1993; Water in the West 1998; Sneddon et al., 2002).

The term "watershed approach" has come to embrace numerous methods for managing water resources that share the goal of harmonizing institutional arrangements for managing water with the natural disposition of surface and groundwater. Regarding surface water, achieving this goal requires a conscientious focus on the drainagebasin as a vehicle for effectively addressing water quantity and quality issues, and for averting and resolving conflicts over interbasin diversion, allocating water for multiple uses, involving the public in decisions, and better managing instream and offstream demands (Ballweber 1995; Shabman 1994; Landre and Knuth 1993; Cortner and Moote 1994). For managing groundwater, other approaches that encourage equity may be needed, such as regional coordination and oversight for other entities that acknowledge the lack of spatial congruence between aquifers and political jurisdictions.

While a number of watershed-based reform efforts are underway—particularly in the United States—to address issues as wide-ranging as stream bank restoration, water quality enhancement, riparian land use planning, local environmental education, critical habitat restoration, source water protection, and a variety of runoff reduction and other water pollution abatement efforts—the verdict is out on how effective these efforts have been, and what factors are needed to enhance their effectiveness (Baer and Pringle 2000; U.S. EPA 1999; McGinnis 1999). One thing that is known is that these efforts have spent considerable time and effort in developing unique and innovative ways of formulating and implementing their plans through community representation mechanisms; public-private partnerships to discourage impervious surface, reduce runoff, enhance filtration, and encourage sustainable land-use development; and other measures.

To chronicle these efforts in a short space—their successes, and limitations—would not pay justice to the challenges they face. Some sense of their diversity is offered by the following geographically diverse examples: New York City, Seattle, Salt Lake City, and several other cities that have purchased land in and around their source water supply watersheds to deter uncontrolled, encroaching development–one model for ensuring the protection of people and vital resources by preventing inequitable encroachments. Since the 1980s, Maryland, Pennsylvania, Virginia, the District of Columbia, Chesapeake Bay Commission, and U.S. EPA have joined together in an oft-cited government-private sector restoration effort to reverse water quality declines and threats to fish and wildlife in the Chesapeake Bay estuary (Williams 2000). Strategies include reductions in point source pollutants, and targeted cooperative and voluntary landowner participation programs to begin addressing nonpoint runoff. The success of this and similar programs is open to contention; as of this writing nitrogen and phosphorous levels remain high in the Chesapeake Bay and water quality in the bay remains low. As the chapters by Kamieniecki and Belows and Wilder illustrate, the devolution of power to subnational, geographically based units does not necessarily insure equitable treatment of disadvantaged individuals and groups.

Adopting an Ethic of Fallibility   Identify policies that are reversible, have low impact, are nonstructural.

Essential to adoption of any system of values designed to promote equity is the need to adopt policies that are self-correcting; that acknowledge, in other words, the fallibility of any policy framework and the need to permit and embrace policy change. Keys to such a system are the following:

• *Inclusiveness and ethical eclecticism*   No ethical approach to the management of water resources should be adopted that categorically excludes any constituency or alternative approach to management. This means that any approach to management should emphasize *process* as much as *substance*—providing a means for the widest possible debate and deliberation over the broadest range of viable, realistic, unbiased alternatives possible. Moreover, there is a consensus among scholars who have tried to apply ethical principles to the analysis of public policy that, regardless of the ethical approach one adopts, some general principles should be adhered to as normative policy objectives. At a minimum, it is not unreasonable to demand of government that water and other public policies be efficient (use natural resources in a beneficial, unwasteful fashion); equitable (be fair and accessible to all who need them, permitting equal sharing of the burdens as well as the benefits of their development); sustainable (be mindful of the interests of future generations); and transparent (lead to policy instruments whose rules and methods are simple and clear).

Practically speaking, though, can ethical eclecticism be reconciled with the need to set priorities among differing goals? The simple answer would seem to be no—the best we can hope to achieve is a just process. But this is too simplistic a solution. The chapters in this volume and other social constructionist literature suggest another possible path. Natural phenomena such as water have different meanings to different groups. These divergent meanings depend on culturally conditioned perspectives that are shaped by social status, a group's goals and aspirations, a society's history, and individual needs (Blatter and Ingram 2001).

We may not be able to reconcile these divergent perspectives, but we can at least aim to make different meanings transparent so that parties to water disputes understand the bases for their disagreement. We can even negotiate the resolution of particular conflicts over water use through mediation of these conflicting perspectives, and through legal reform to permit the equitable representation of these perspectives. Finally, we can show how noninstrumental uses of water are legitimate and important to some groups—as illustrated in the actions of the residents in the San Luis Valley whose motivations described by Clay Arnold were to preserve their culture and way of life for future generations.

• *Commitment to Democratic decision making*   Diverse, even divergent ethical approaches to the environment can, if filtered through the appropriate political frameworks, serve as complementary rather than mutually exclusive prisms for viewing nature, the world, and our obligations to both. To make this ethical pluralism possible, and to derive benefits from the exchange of views these different approaches provide, we must demand that decision-making institutions be responsive to deliberative processes of discussion, reflection, the weighing of alternatives, and justification for choosing particular policies—and be willing to justify these policies by reference to the broader needs of a society, not a single region (Chrislip and Larson 1994).

The public must be assured that water resource decisions are not—as they have too often been in the past—preformed or predecided. Moreover, ethical inclusiveness implies accepting responsibility for unanticipated impacts through embracing the use of consultative mechanisms in decision making. Decision makers must acknowledge that scientific reasoning is often probabilistic in nature, and must be aware that people who are "motivated to deceive either themselves or others about an issue may use forms of reasoning . . . that contain hidden fallacies" (Ellis 1998, 179).

We can approximate a setting of priorities if we acknowledge these reasoning fallacies and thereby the need for scientifically supported policy propositions that are based on the best available evidence. What would qualify as evidence? At a minimum, propositions would be predictive, statistically significant, contain well-defined terms based on observable information, and would eliminate plausible alternative explanations for what is causing a particular water resource problem (Ellis 1998, 179–180). In many cases in this volume, the jaundiced or prejudicial use of scientific evidence to justify a political decision over the management of water has time and again been used to thwart reform, or reformers.

• *Making ethical assumptions clear and transparent*   No ethical approach will please everyone nor achieve universal acclaim. However, to the extent that we articulate our ethical claims publicly, in full view of others and other approaches—not in some secluded bureaucratic environment that lacks transparency and breeds both interpersonal and governmental distrust (Warren 1999)—and so long as we work to ensure that others understand our assumptions and try to understand theirs—then we stand a better chance of ensuring that whatever perspectives we adopt conform to a broader notion of justice—justice, that is, as fairness (e.g., Rawls 1971). Such an approach is modest, cautious, and careful. It would advocate no major change in the status quo designed to make any group or region better off at the expense of making another region or group worse off.

• *Collaborating with those with whom you disagree*   For quite some time, much of the pioneering work in collaborative resource management has been arguing for the need for collaborative learning. The idea is that genuine collaboration that permits the putting-aside of self-interest and partiality requires the development of a collaborative stakeholder process that permits independent, role-bound adults to be able to engage in a safe, secure, nonthreatening, open process of cooperation based on mutual respect, integrity, and actual teamwork in the development of solutions (Daniels and Walker 2001). Collaboration overcomes resistance to change, facilitates new opportunities for funding, and stimulates resilient policy ideas.[8]

Second, these criteria are benchmarks that may help identify deficiencies in current decision-making frameworks. In essence, they can be used to gauge whether a proposed decision is likely to benefit an entire region or only a part of it, enhance or impede the ability to form partnerships for resolving water problems, and produce durable decisions acceptable to stakeholders. As many of the cases in this volume have shown, such benchmarks are implicitly important in disputes involving regional water resources, because they embrace both substantive ends and the procedural means to attain them, ensure that they are maintained, and make possible their reform.

Further Research Needs

A great many challenges regarding equity and water remain to be explored–particularly as regarding how to measure and evaluate successful implementation of schemes intended to redress inequities. While this volume acknowledges the importance of granting to equity concerns a far greater role in water policymaking than has traditionally been the case, as these chapters illustrate, achieving equity is difficult and often arduous. A number of issues that require further research include:

• Are there best practices that can be chronicled, described, and, perhaps, compiled into a permanent network that can be shared among NGOs, public officials, scholars, and the general public on equitable practices that work well, and whose lessons can be adapted by others? This network could serve as a permanent clearinghouse to gather, compile, and update cases and other examples of success. At the same time, context and place are important in water resources decision making, and successes may not be portable, especially to localities without long histories of collaborative decision making.

• Regarding the implementation of equity practices, institutions, policies, and procedures, what do we know about the conditions for bringing these conditions about—and promoting their adoption—that can be employed to leverage the talents and resources of international aid organizations and other NGOs?

• What are the respective—and possibly divergent roles—of markets and political institutions in contributing to, and overcoming, inequities in the allocation, distribution, and governance of water? As many of the contributors to this volume have noted, there remains considerable debate over the problem of fragmented water institutions, the need for better policy coordination and open decision arenas on the one hand, with growing pressures to privatize water (and presumably hasten market-driven investments that might provide desperately needed new or expanded infrastructure) on the other. Questions about the relative roles of markets, government, and public and private partnerships deserve further consideration.

While growing privatization of water infrastructure worldwide prompts questions about *justice* by directing our attention to charges that private ownership leads to monopolistic control of water, unfair allocation, and limited public access to decision making, we also know that historically, policy failures regarding water equity have as often been the result of misguided government decisions to dam rivers, divert water, and buy and sell water rights in order to move water from uses deemed low in value to those deemed to be more advantageous to the interests of the rich and powerful (e.g., Shiva 2002; Barlow and Clarke 2002; Ascher 1999).

In short, there is no dearth of important research and policy questions that remain to be addressed. The contributors to this volume have boldly identified a number of challenges and, through carefully drawn cases, depicted the ways that various combinations of poverty, underdevelopment, chronic water shortage, weak legal capacity, and transboundary conflicts between wealthy and powerful—and disadvantaged and weak—communities challenge our intuitive notions of fairness and justice in water resources management. While much work is needed, these contributions have blazed a positive path.

Notes

1. The theologian and political philosopher Reinhold Niebuhr advanced one of the best formulations of this problem; see *Moral Man and Immoral Society* (New

York: Simon and Schuster, 1932; reprinted 1960). Niebuhr urges: "If conscience and reason can be insinuated into the resulting struggle (of politics) they can only qualify but not abolish it" (p. xiii).

2. See, for example, Deuteronomy 4:5–6: "See, I have taught you decrees and laws as the LORD my God commanded me, so that you may follow them in the land you are entering to take possession of it. Observe them carefully, for this will show your wisdom and understanding to the nations, who will hear about all these decrees and say, 'Surely this great nation is a wise and understanding people.'"

3. In Genesis 9, God blessed Noah and his descendants by stating: "I now establish my covenant with you and with your descendants after you and with every living creature that was with you—the birds, the livestock and all the wild animals, all those that came out of the ark with you, every living creature on earth. . . . never again will the waters of a flood cut off all life; never again will there be a flood to destroy the earth. And God said, This is the sign of the covenant I am making between me and you and every living creature with you, a covenant for all generations to come: I have set my rainbow in the clouds, and it will be the sign of the covenant between me and the earth. Whenever I bring clouds over the earth and the rainbow appears in the clouds, I will remember my covenant between me and you and all living creatures of every kind. Never again will the waters become a flood to destroy all life. Whenever the rainbow appears in the clouds, I will see it and remember the everlasting covenant between God and all living creatures of every kind on the earth" (NIV: 9–16).

This reading was a reinvocation of a perpetual covenant, as seen in Deuteronomy 7: 11–15: "Therefore, take care to follow the commands, decrees and laws I give you today. If you pay attention to these laws and are careful to follow them, then the LORD your God will keep his covenant of love with you, as he swore to your forefathers. He will love you and bless you and increase your numbers. He will bless the fruit of your womb, the crops of your land—your grain, new wine and oil—the calves of your herds and the lambs of your flocks in the land that he swore to your forefathers to give you. You will be blessed more than any other people; none of your men or women will be childless, nor any of your livestock without young. The LORD will keep you free from every disease. He will not inflict on you the horrible diseases you knew in Egypt, but he will inflict them on all who hate you."

4. Hobbes's distinction between "covenant," an agreement based on trust between parties for performance of an act at some indeterminate time, and "contract," which was viewed as a "mutual transference of right," is also instructive. Clearly, he appreciated his readers' familiarity with biblical language and deliberated on this distinction as a logical departure from his views on religion,which saw the emergence of God as "created by human fear" (see *Leviathan Parts I and II*, ed. Herbert W. Schneider [Indianapolis: Bobbs-Merrill, 1958]: 94, 112).

5. See, for example, Romans 8: 19 ff: "The creation waits in eager expectation for the sons of God to be revealed. For the creation was subjected to frustration, not by its own choice, but by the will of the one who subjected it, in hope that

the creation itself will be liberated from its bondage to decay and brought into the glorious freedom of the children of God." Note, also, that water baptism—a symbol of liberation from the oppression of sin that separates us from God—is a water-centered ritual. It has its origins in the symbolism of the Israelites being led by Moses out of slavery in Egypt through the Red Sea and from the baptism of Jesus by John the Baptist in the Jordan. After Jesus' resurrection, he commanded his disciples to baptize "in the name of the Father, Son, and Holy Spirit" (Matthew 28:19–20).

6. Or, in New Covenant terms: "So in everything, do to others what you would have them do to you, for this sums up the Law and the Prophets" (see: New Testament, Book of Matthew 7:12).

7. Recent scholarship has suggested that Powell's environmental views of progressive conservation owed at least as much to his Methodist upbringing as to his belief in utilitarian ethics. As Donald Worster notes, he was named for his parents' hero, John Wesley, and "would grow up, they prayed, to become a spiritual leader for America." Furthermore, he was taught to "put the love of God above the things of the earth and subdue his naturally selfish will" (*A River Running West: The Life of John Wesley Powell* [New York: Oxford University Press, 2001], 5). For another view of Powell and his values, see Stephen J. Pyne, *How the Canyon Became Grand: A Short History* (New York: Penguin, 1999).

8. The theologian and political philosopher Reinhold Niebuhr advanced one of the best formulations of this problem; see *Moral Man and Immoral Society* (New York: Simon and Schuster, 1932; reprinted 1960). Niebuhr urges: "If conscience and reason can be insinuated into the resulting struggle (of politics) they can only qualify but not abolish it" (p. xiii).

References

Adams, Richard E., and Dannele E. Peck. 2006. "Climate Change and Water Resources: Potential Impacts and Implications," paper presented to the 4th Rosenberg International Water Forum, Calgary. http://Rosenberg.ucanr.org.

Anderson, Charles W. 1979. "The Place of Principles in Policy Analysis." *American Political Science Review* 73 (September): 711–723.

Ascher, William. 1999. *Why Governments Waste Natural Resources: Policy Failures in Developing Countries*. Baltimore: The Johns Hopkins University Press.

Baer, K. E., and C. M. Pringle. 2000. "Special Problems of Urban River Conservation: The Encroaching Megalopolis" In P.J. Boon, B.R. Davies, and G.E. Potts, eds., *Global Perspectives on River Conservation: Science, Policy and Practice*. New York: John Wiley and Sons, pp. 385–402.

Ballweber, J. A. 1995. "Prospects for Comprehensive, Integrated Watershed Management under Existing Law." *Water Resources Update* 100 (Summer): 19–27.

Barlow, Maude, and Tony Clarke. 2002. *Blue Gold: The Fight to Stop the Corporate Theft of the World's Water*. New York: The New Press.

Baum, Ellen. 2004. *Wounded Waters: the Hidden Side of Power Plant Pollution.* Boston, MA: Clean Air Task Force, February.

Beatley, Timothy. 1994. *Ethical Land Use: Principles of Policy and Planning.* Baltimore: Johns Hopkins University Press.

Bellah, Robert N., ed. 1996. *Habits of the Heart: Individualism and Commitment in American Life.* Berkeley: University of California Press.

Blatter, Joachim, and Helen Ingram, eds. 2001. *Reflections on Water: New Approaches to Transboundary Conflicts and Cooperation.* Cambridge, MA: MIT Press.

Blatter, Joachim, Helen Ingram, and Suzanne Levesque. 2001. "Expanding Perspectives on Transboundary Water." In Joachim Blatter and Helen Ingram, eds., *Reflections on Water: New Approaches to Transboundary Conflicts and Cooperation.* Cambridge, MA: MIT Press, pp. 31–53.

Blumm, Michael. 1998. "Columbia River Basin." In Robert E. Beck, ed., *Waters and Water Rights vol. 6,* Charlottesville, VA: Michie Publishing).

Bright, John. 2000. *A History of Israel.* Westminster: John Knox Press.

Burton, Lloyd. 1991. *American Indian Water Rights and the Limits of Law.* Lawrence: University Press of Kansas.

Butler, Lynda L. 1990. Butler, "Environmental Water Rights: An Evolving Concept of Public Property." *Virginia Environmental Law Journal 9,* no. 2 (Spring): 323–379.

Caney, Simon. 2001. "Review Article: International Distributive Justice," *Political Studies 49,* no. 5 (December): 974–997.

Chrislip, David D., and Carl E. Larson. 1994. *Collaborative Leadership: How Citizens and Civic Leaders Can Make a Difference.* San Francisco: Jossey-Bass.

Conca, Ken. 2006. *Governing Water: Contentious Transnational Politics and Global Institution Building.* Cambridge, MA: MIT Press.

Conniff, Richard. 2007. "Third World to Bear Brunt of Global Warming." *Environment: Yale,* Spring, pp. 27–29.

Cortner, Hanna A., and Margaret A. Moote. 1994. "Setting the Political Agenda: Paradigmatic Shifts in Land and Water Policy." In R. Edward Grumbine, ed., *Environmental Policy and Biodiversity,* Washington, DC: Island Press, pp. 365–377.

Daniels, Steven E., and Gregg B. Walker. 2001. *Working through Environmental Conflict: The Collaborative Learning Approach.* Westport, CT: Praeger Publishers.

DeWitt, Calvin. 2002. *Remarks on Stewardship, Ecology, Theology and Judeo-Christian Environmental Ethics: A Conference at the University of Notre Dame,* February, 21–24.

Dugger, Celia W. 2007. "As Prices Soar, U.S. Food Aid Is Buying Less" *New York Times,* September 29.

Ellis, Ralph D. 1998. *Just Results: Ethical Foundations for Policy Analysis.* Washington, DC: Georgetown University Press.

Feldman, D. 2007. *Water Policy for Sustainable Development*. Baltimore: Johns Hopkins University Press.

FO Licht's World Ethanol and Biofuels Report, 2007. Thursday, August 17.

Fowler, Robert Booth. 1995. *The Greening of Protestant Thought*. Chapel Hill: University of North Carolina Press.

Gelt, Joe. 1997. "Sharing Colorado River Water: History, Public Policy, and the Colorado River Compact." *Arroyo* 10 (August).

Gleick, P. H. 2000. *Water: the Potential Consequences of Climate Variability and Change for the Water Resources of the United States*. A report of the National Water Assessment Group for the U.S. Global Change Research Program. Pacific Institute for Studies in Development, Environment, and Security, Oakland, CA.

Graham-Rowe, Duncan. 2005. "Hydroelectric power's Dirty Secret Revealed," *New Scientist* 29 (February 14): 24.

Grizzle, Raymond E., and Christopher B. Barrett. 1996. *The One Body of Christian Environmentalism*. UAES Journal Paper, Pew Charitable Trust Report, August. http://cesc.montreat.edu/GSI/GSI-Conf/Mini-Grants/Taylor-OneBody.html

Hartig, J. H., et al. 1992. "Identifying the Critical Path and Building Coalitions for Restoring Degraded Areas of the Great Lakes." In *Water Resources Planning and Management: Saving a Threatened Resource*. New York: Conference on Water Resources Planning and Management, ASCE.

Hutson, Susan S., et al. 2005. "Estimated Use of Water in the United States in 2000." US Geological Survey, http://pubs.usgs.gov/circ/2004/circ1268/

Ingram, Helen. 2001. "Falling on Deaf Ears? Science and Natural Resources Policy." Working Paper, University of California–Irvine.

IPCC. 2007a. *Climate Change 2007: The Physical Science Basis. Contribution of Working Group I to the Fourth Assessment Report of the Intergovernmental Panel on Climate Change*. Solomon, S., et al., eds. Cambridge, UK: Cambridge University Press.

IPCC. 2007b. *Climate Change 2007: Impacts, Adaptation, and Vulnerability. Working Group II Contribution to the Intergovernmental Panel on Climate Change Fourth Assessment Report. Summary for Policymakers*. Neil Adge, et al. Geneva, Switzerland: IPCC Secretariat, April.

International Water Management Institute, Stockholm. 2007. *SciDev.net*. http://www.iwmi.cgiar

Kant, Immanuel. 1975 edition. *Foundations of the Metaphysics of Morals*. Translated with an introduction by Lewis Beck. Indianapolis, IN: Bobbs-Merrill.

Kantrowitz, Arthur. 1993. "Elitism vs. Checks and Balances in Communicating Risk Information to the Public." *Risk, Health, Safety, and the Environment* 4: 101.

Kavka, Greg. 1978. "The Futurity Problem." In R. I. Sikora and Brian Berry, eds., *Obligations to Future Generations*. Philadelphia: Temple University Press.

Laitos, Jan G., and Joseph P. Tomain. 1992. *Energy and Natural Resources Law in a Nutshell.* St. Paul, MN: West Publishing.

Landre, B. K., and B. A. Knuth. 1993. "Success of Citizen Advisory Committees in Consensus Based Water Resources Planning in the Great Lakes Basin." *Society & Natural Resources* 6 (3): 229.

Landry, Clay J. 2000. "Agriculture and Water Markets in the New Millennium." *Water Resources Impact* 2, no. 3 (May): 13–15.

Lasswell, Harold D. 1958. *Politics: Who Gets What, When and How?* New York: Peter Smith.

Lee, Kai N. 1993. *Compass and Gyroscope: Integrating Science and Politics for the Environment.* Washington, DC: Island Press.

Leopold, Aldo. 1949. *A Sand County Almanac: With Essays on Conservation from Round River.* New York: Oxford University Press.

McFague, Sally. 1993. *The Body of God: An Ecological Theology.* Minneapolis: Fortress Press.

McGinnis, Michael V. 1999. "Making the Watershed Connection." *Policy Studies Journal* 27 (3): 497–501.

Merideth, Robert. 2001. *A Primer on Climatic Variability and Change in the Southwest.* Tucson: Udall Center for Studies in Public Policy & Institute for the Study of Planet Earth, University of Arizona, March.

Moody, Tom. 1997. "Glen Canyon Dam: Coming to an Informed Decision." *Colorado River Advocate—Grand Canyon Trust* (fall). http://www.glencanyon .org/Articles97.html

Niebuhr, Reinhold. 1932/1960. *Moral Man and Immoral Society.* New York: Simon and Schuster.

Pearson, Rita P. 2000. "Managing Water Scarcity—Southwestern Style." *Troubled Waters—Managing a Vital Resources—Global Issues Electronic Journal,* March.

Quintana, Victor M. 2007. "Biofuels and Small Farmers." *Americas Program,* August 26. http://americas.irc-online.org/am/4510

Rawls, John. 1971. *A Theory of Justice.* Cambridge: Belknap Press of Harvard University Press.

Restoring the Waters. 1997. Boulder, CO: Natural Resources Law Center, the University of Colorado School of Law, May.

Rogers, Peter P. 2006. "Water Policy and Management." In Peter Rogers, M. Ramon Llamas, and Luis Martinez-Cortina, eds., *Water Crisis: Myth of Reality.* London: Taylor and Francis, pp. 3–36.

Rolston, Holmes. 1988. *Environmental Ethics: Duties to and Values in the Natural World.* Philadelphia: Temple University Press.

Runge, C. Ford, and Benjamin Senauer. 2007. "How Biofuels Could Starve the Poor." *Foreign Affairs* 6, no. 3 (May/June): 41–53.

Sachs, Wolfgang, ed. 2002. "The Jo'berg Memo: Fairness in a Fragile World." Berlin Heinrich Boll Foundation. http://www.joburgmemo.org

Selbourne, John. 2000. *The Ethics of Freshwater Use: A Survey.* Paris: UNESCO, 2.

Shabman, L. 1994. "Bargaining, Markets, and Watershed Restoration: Some Elements of a New National Water Policy." In M. Reuss, ed. *Water Resources Administration in the United States: Policy, Practice, and Emerging Issues.* East Lansing: Michigan State University Press, pp. 94–104.

Shiva, Vandana. 2002. *Water Wars: Privatization, Pollution, and Profit.* London: Pluto Press.

Sneddon, Chris, et al. 2002. "Contested Waters: Conflict, Scale and Sustainability in Aquatic Socioecological Systems." *Society & Natural Resources* 15, no. 8 (September): 663–675.

U.S. Environmental Protection Agency. 1999. *Protecting Sources of Drinking Water: Selected Case Studies in Watershed Management.* EPA 816-R-98-019. Washington, DC: Office of Water, EPA, April.

Volkman, John M. 1996. *A River in Common: The Columbia River, the Salmon Ecosystem, and Water Quality.* Denver, CO: Western Water Policy Review Advisory Commission.

Warren, Mark E. 1999. "Conclusion." In Mark E. Warren, ed., *Democracy and Trust.* Cambridge, UK: Cambridge University Press, pp. 346–360.

Water in the West: Challenge for the Next Century. 1998. Report of the Western Water Policy Review Advisory Commission. Springfield, VA: National Technical Information Service, June.

Williams, John Page. 2000. *Another Side of the TMDL Discussion.* Chesapeake Notebook, 2-2000. http://www.cbf.org

World Commission of Dams. 2000. *Dams and Development: A New Framework for Decision Making. The Report of the World Commission on Dams.* London: Earthscan.

Worldwatch Institute. 2007. *Biofuels for Transport: Global Potential and Implications for Energy and Agriculture.* London: Earthscan, p. 206.

Contributors

T. Clay Arnold is Professor of Political Science at the University of Central Arkansas. His scholarship features western water politics and issues in contemporary social and political theory. He has published articles and essays in the *American Political Science Review, Politics & Policy, the Journal of the Southwest,* and the SUNY Press series in Environmental Philosophy and Ethics.

Madeline Baer is a graduate student in the Department of Political Science at the University of California, Irvine. She received her B.A. in Anthropology from American University, and her M.A. in political science from the University of California, Irvine. Her interests include human rights, the impact of neoliberal policies on development, and the struggle for water rights in Latin America.

Amy Below earned her B.A. in Environmental Studies from the University of California at Santa Barbara and her M.A. in Environmental Studies with an emphasis in Law, Politics and Public Policy from the University of Southern California. She is currently a doctoral student at the University of Southern California in the Politics and International Relations Program. Her current research interests include comparative politics, environmental policy, and foreign policy analysis, with specific attention to environmental foreign policy decision making.

David Lewis Feldman is professor and chair of the Department of Planning, Policy, and Design at the University of California, Irvine. He was Professor and Head of the Department of Political Science at the University of Tennessee. Before that, he was a researcher at University of Tennessee's Energy, Environment and Resources Center and a research staff member at Oak Ridge National Laboratory. His research focuses on water resources management, global climate change, and resource disputes. His most recent book, published by Johns Hopkins University Press in 2007, is titled *Water Policy for Sustainable Development.* He is the author of four other books on water, energy, and climate change, and more than sixty articles. Feldman is the 2001 recipient of the Policy Studies Organization Interdisciplinary Scholar Award and served as editor of *The Review of Policy Research* and symposium coordinator for the *Policy Studies Journal.* His current research explores the relationship between the growth of a democratic civil society and environmental reform in Russia.

_t_ is Associate Professor of History at Arizona State University. He spe-
_n_ the history of the American West, public lands policy, and environmen-
ry. He is the author of the book _A Conspiracy of Optimism: Management_
_ᴜ_ⱼ _˖˖_ _lational Forests since World War Two_ (Nebraska 1994), and the editor of
two anthologies on the environmental history of the Pacific Northwest, including
Northwest Lands, Northwest Peoples: Readings in Environmental History (University of Washington 1999), coedited with Dale Goble. He is currently writing a
book on the history of electric power in the Pacific Northwest. In addition to his
scholarship, Hirt has worked with a variety of conservation groups promoting
biodiversity and wildlands conservation in the American West, and he currently
serves as president of the board of directors of the Sky Island Alliance.

Helen Ingram is Research Fellow at the Southwest Center at the University of
Arizona. She is professor emeritus at both the University of Arizona and the
University of California at Irvine. She began her interest in water and equity in
1988 when she coauthored _Water and Poverty in the Southwest_. Tucson, AZ:
University of Arizona Press, 1987. She coedited a previous book in this MIT Press
series, _Reflections on Water: New Approaches to Transboundary Conflicts and_
Cooperation (2001). Formerly she was Warmington Endowed Chair in Social
Ecology at the University of California at Irvine and Director of the Udall Center
at the University of Arizona.

Sheldon Kamieniecki is Dean of the Division of Social Sciences at the University
of California, Santa Cruz. He has published several books and numerous journal
articles and book chapters on local, state, federal, and international environmental policy issues. He is the recipient of the Lynton K. Caldwell Prize for the best
book published between 1995 and 1997 and for the best book published between
2004 and 2006 on environmental politics and policy from the Section on Science,
Technology and Environmental Policy in the American Political Science Association. He recently coauthored a book on strategic planning in environmental
regulation for the MIT Press, published a second book on the influence of business over environmental policy with the Stanford University Press, and coedited a
third book on the influence of business over environmental policy with the MIT
Press. He is the coeditor of a book series on American and Comparative Environmental Policy at the MIT Press.

Maria Carmen Lemos is an associate professor of natural resources and environment at the University of Michigan and a senior policy analyst at the Udall
Center for Studies of Public Policy at the University of Arizona. From 2006–2007
she was a James Martin Fellow at the Environmental Change Institute at Oxford University. Her research interests focus on the human dimensions of global
climate change—especially concerning the use of technoscientific knowledge in
climate-related policy and adaptation in less developed countries; the impact of
technocratic decision making on democracy and equity; and natural resources
governance (especially water) and climate.

Stephen Mumme is Professor of Political Science in the Department of Political
Science at Colorado State University. His research effort centers on water and environmental management along the U.S.-Mexican border and within the North

American region. He is coauthor with Alan Lamborn of *Statecraft, Domestic Politics, and Foreign Policymaking: the El Chamizal Dispute* (1988) and author or coauthor of numerous articles on institutional aspects of water management on the U.S.-Mexican border. His most recent article, coauthored with Donna Lybecker, "The All-American Canal: Perspectives on the Possibility of a Bilateral Agreement," appears in Vicente Sanchez Munguia, ed., *Lining the All-American Canal: Competition or Cooperation for Water in the U.S.-Mexico Border?* San Diego: SCERP Monograph Series.

Richard Warren Perry is Professor of Justice Studies at San Jose State University. He works on legal theory and cultural anthropological studies of law, governmentality, and globalization. His most recent publications have addressed questions of environmental justice and the status of indigenous peoples under expanding transnational market regimes.

Ismael Vaccaro is Assistant Professor at McGill University (Department of Anthropology and McGill School of Environment). His work, currently focused on the Pyrenees and Mexico, deals with landscape analysis and the impact of conservation policies on local communities and their access to natural resources. His recent research and publications are analyzing the emergence of new uses of the territory characterized by a patrimonialization and urban consumption of nature and culture.

John M. Whiteley is Professor of Social Ecology at the University of California, Irvine. His most recent scholarship involves the Quest for Peace in the Nuclear Age. This is an initiative available in digital form from the UCI Libraries on the internet. He is the coauthor (with Russell J. Dalton, Paula Garb, Nicholas P. Lovrich, and John C. Pierce) of *Critical Masses: Citizens, Nuclear Weapon Production, and Environmental Destruction in the United States and Russia* (MIT Press).

Margaret Wilder is Assistant Professor of Latin American Studies and Geography at the University of Arizona. She holds a bachelor's degree in Government and International Affairs, University of Notre Dame; a master's degree in Public Policy Studies, University of Chicago; and a Ph.D. in Geography, University of Arizona. Dr. Wilder's research is focused on environment and poverty/development issues in Latin America. She is interested in the political ecology of water in Mexico, and has done extensive research in the areas of water policy reforms, new water management institutions, and the implications of these for equity and sustainability. From a theoretical perspective, she is interested in the transformation of state-society relations under neoliberal regimes of governance. Dr. Wilder also works on environmental sustainability on the Mexico-U.S. border, and currently has a stakeholder-based conservation study in the Lower Colorado River and Delta binational region. Dr. Wilder is a co-PI on the Climate Assessment of the Southwest Project at the University of Arizona and was a Udall Center Fellow at the Udall Center for Public Policy Studies in 2006–2007.

Index